Computer Integrated Planning
and Design for Construction

A. Retik & D. Langford

┱┓Ⅰ Thomas Telford

Published by Thomas Telford Publishing, Thomas Telford Ltd, 1 Heron Quay, London E14 4JD.

URL: http://www.thomastelford.com

Distributors for Thomas Telford books are
USA: ASCE Press, 1801 Alexander Bell Drive, Reston, VA 20191-4400, USA
Japan: Maruzen Co. Ltd, Book Department, 3 – 10 Nihonbashi 2-chome, Chuo-ku, Tokyo 103
Australia: DA Books and Journals, 648 Whitehorse Road, Mitcham 3132, Victoria

First published 2001

Also available from Thomas Telford Books

Information Technology in construction design. Michael Phiri. ISBN 07277 2673 0
Infoculture. Stephen Vincent and Scott Wilson Kirkpatrick. ISBN 0 7277 2597 1
Interdisciplinary design in practice. Edited by Sebastian Macmillan, Paul Kirby & Robin Spence. ISBN 0 7277 3008 8

A catalogue record for this book is available from the British Library

ISBN: 0 7277 3007 X

Typeset by MHL Typesetting Ltd, Coventry
Printed and bound in Great Britain by MPG Books, Bodmin

To my parents

A. Retik

Acknowledgements

The authors are indebted to many people who assisted in the writing and production of this book. Our ideas have been greatly informed by lively discussions with colleagues at Strathclyde University. Bimal Kumar, John Conlin and Iain MacLeod have been particularly influential on our thinking. In a wider world Aviad Shapira from Technion, Israel a Visiting Scholar to Strathclyde made incisive contributions at difficult stages in the writing of the book.

For the graphics we would like to thank Nora Retik, Stuart Buchanan and Andrew Layden, and for the production of the text Marianne Halforty and Bernadette Cairns.

Contents

Acknowledgements iv
Introduction vii

Part 1 Background to Information Technology 1

Chapter 1 Computer Equipment and Communications 3
Chapter 2 Databases and Computer Graphics 17
Chapter 3 Technical Applications of Computer Software 33
Chapter 4 Computerised Organisation and the Use of Information 52

Part 2 Applications of IT to Construction Projects 71

Chapter 5 Computer-aided Project Planning and Scheduling 73
Chapter 6 Computer-integrated Building Representation for Design and Construction 104
Chapter 7 Computer-integrated Multidisciplinary Concurrent Engineering 112

Part 3 Applications of IT in the Construction Business 123

Chapter 8 The Strategic Use of Information Technology 125
Chapter 9 An Application of IT to Strategy Formulation in Construction Firms 145
Chapter 10 Conclusions 157

Appendix A Cases of IT Implementation: Evolution or Revolution? 159
Appendix B Cases of IT Implementation Using the Internet 173
CAD Glossary 177

Index 185

Introduction

'The computing industry is changing so fast, nobody can know what it will be like in 10 or 15 years' time.'
Andy Grove, Chief Executive of Intel (*Sunday Times*, 9 February 1997)

The construction industry does not change as fast as the computer industry. This results often in an inertia (some would say conservatism) in all construction-related areas, including educational institutions. As a result, today's construction and civil engineering education systems do not always properly equip graduates with the skills required to operate modern information technology (IT). Moreover, the high pace of IT development makes it very difficult for practising engineers and managers to keep up with a wide range of novel tools and approaches. Also, many senior managerial staff often lack formal broad education in computing or information sciences, making it difficult for them to follow trends and cope with strategic decision making.

On the other hand, information explosion, globalisation, international competition and new procurement techniques demand solutions that can only be achieved by proper information management. Harnessing IT to take a competitive advantage is a crucial goal and vital task for many organisations.

This book is about the application of advanced information technology to planning and design in construction. The book not only describes and explains the current applications of computer tools, but also presents new ideas for the new use of these tools in planning and design processes, concentrating on preliminary stages of the construction process. The main aim of this text is to demonstrate the implementation of these ideas and uncover many extraordinary opportunities for construction process improvements. An important feature of this book is its relevance to the work of all participants in design and construction of buildings – from architects and engineers to project managers. We use clear engineering language to describe the computer science behind the information technology applications that are featured in the text.

Part 1 gives a background to information technology. Chapters 1–3 explore the background to current developments in IT, starting with a brief overview of the existing 'traditional' computerised tools and approaches, and then developing the discussion to focus on advanced tools such as communications technologies (Internet and videoconferencing), advanced visualisation tools (such as geographical information systems and virtual reality), artificial intelligence and related subjects (knowledge-based system, object orientation and neural networks). Examples of existing applications are presented before new developments in research and industrial applications are discussed.

Chapter 4 presents the fundamentals of information science. Terms such as data, information and knowledge are defined. Then the systems approach and management information system concepts are introduced, described and demonstrated. Lastly, information sources (both online and offline computerised databanks and traditional institutional and organised sources) that could be used to support planning and design are outlined.

The three chapters in Part 2 look at the application of IT to construction projects. Chapter 5 is about computer-based planning and management of projects. It starts with a review and analysis of existing planning and scheduling tools. It then depicts novel and advanced tools, such as telecommunications and visual simulation to support integration of planning and design information. Examples of recent research involving construction companies are also provided.

Chapter 6 analyses the data and information needed for integrated planning and design and shows the way it can be presented and modelled.

Chapter 7 presents and demonstrates a comprehensive approach to design and planning, emphasising their multi-disciplinary character. It also deals with the practical aspects of computerisation and application development and illustrates the above approach describing a knowledge-based system prototype.

Part 3 looks at the application of IT in the construction business. The two chapters making up this part are dedicated to the strategic use of information technology at the company level. Chapter 8 delves into how information and IT may be yoked to the business strategy of construction companies. It links investment in IT and how this can deliver competitive advantage to forward-looking companies. Issues such as using IT to change the business process for construction companies are featured in this chapter.

Chapter 9 presents a case of how a particular IT application has been used to inform the development of strategic planning for construction companies. The chapter reports the findings of research undertaken at Strathclyde University in which the business environment for the construction industry was mapped and how the components of this business environment were linked together. The IT tool developed was called CAFE (Construction Alternative Future Explorer). The system can create complex interlocking pictures of the environment, which frees personnel to think creatively about how they can respond to the opportunities or threats that present themselves while seeking to explore the future business environment for construction.

The eight case studies in the two appendices are examples of successful IT implementation. The descriptions focus on the implementation process and decision making within the companies.

The book concludes with a CAD glossary.

This book will provide readers with an overview of modern computing tools and equip them with independent judgement and decision making about how IT can be applied to construction settings.

Part 1

Background to Information Technology

Chapter 1

Computer Equipment and Communications

Introduction

The purpose of this first part of the book is to arm readers with a basic knowledge of information technology. Terminology is introduced and explained in simple language and each section is self-contained.

Hardware

The pace of progress in computer technology is one of the twentieth century's most important and existing phenomena. A birthday card that plays a tune when it is opened contains more computing power than existed in the world in 1950. The growth of the industry has spanned global networks served by multi-national corporations. The wealth of a nation depends upon the success of its computer-based communications technology, global networks and the management of information.

Not only is computing power growing exponentially, but an important characteristic of this phenomenon is the continual reduction in the costs of the equipment. Where could it lead? In our view, within a few years, the cost of a personal computer (PC) will be around that of today's TV set. There will be two basic configurations: a stationary computer and a mobile computer. The stationary one will range from a simple network computer or NC (available today – see Figure 1.1) to a scientific computer (SC) which will be similar to today's PC. A mobile computer (MC) will be a mixture of portable computers and mobile phones. Once a 'virtual screen' is invented, the size will not matter. Another development is a shift towards speech rather than keyboard based input and operations. This, coupled with advances in reducing size of the microprocessors (nano-technology is one of the examples), is being reflected in so called 'embedded' computing and 'smart' devices. The range of products can vary from home appliances (i.e. refrigerator that can send a delivery order to a designated shop once the level of beer is below par), to construction components or materials (i.e. a special sensor attached to a

PC/SC/Workstation

Camera

Plotter

Keyboard

Mouse

Printer

Peripheral equipment

Tablet/digitiser

ISP (Internet service providers)

Applications service providers

Network/internet

NC−network computer

Peripheral equipment

Peripheral equipment

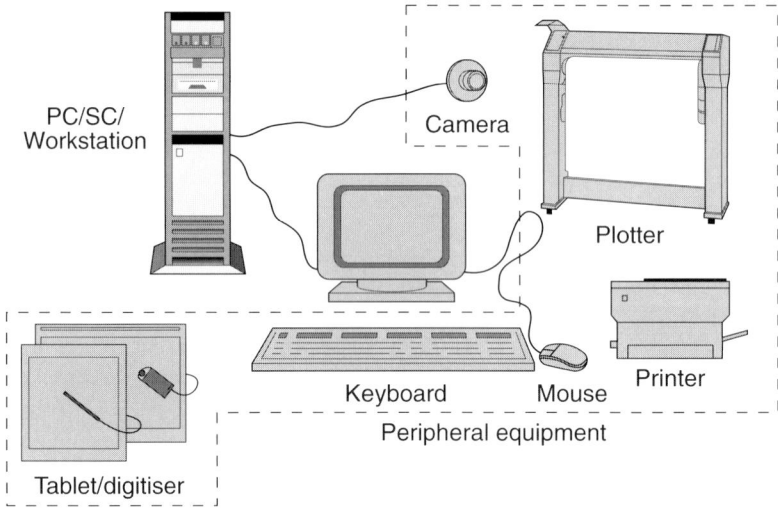

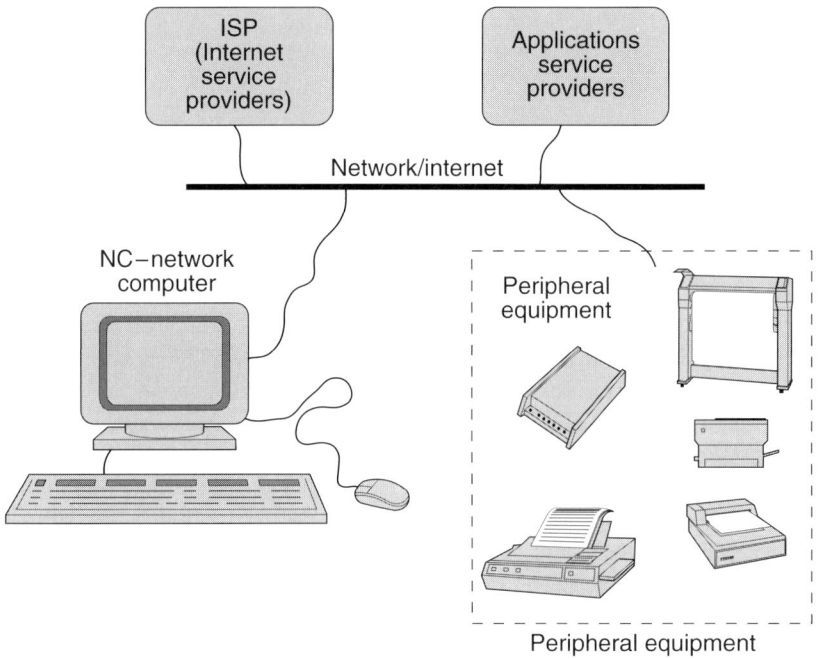

Fig. 1.1   Schematic computer configurations

reinforcement bar within a concrete deck that can inform when the level of corrosion or tension are above the allowed). Such devices can be wirelessly connected to a network thus allowing one to access, monitor and operate them from virtually any place.

However, if the situation with hardware seems to be clear and going in the direction that almost 'everyone wants', the same cannot be said about the situation with software. Software development and programming tools still require quite a heavy human involvement. This results either in too many commercial computer applications produced to solve the same problem (for example, more than 200 project management and scheduling software packages are available; see Aouad and Price, 1994), or in companies developing their own tools (for example, for a company's finance and cost management). In either case the balance of cost and quality is difficult to maintain.

With an increase in 'hardware independent' software tools and advances in Internet technologies, software issues such as development automation, user-oriented operation and maintenance, integration and others are more in the focus of the research and development communities. This is also the case for the construction industry and, therefore, will be of more interest to readers.

Computer Software, Programs and Applications

Whilst most readers will be familiar with the basic terminology it may be helpful to briefly introduce the key elements of IT.

Operating Systems

The operating system starts your computer running and controls the operation of its activities. It manages the entry, flow and display of software and data to and from each part of your computer system. To run a program, you first need to run the operating system.

Any operating system manages the flow of information and performs (with or without a user) the following tasks:

- manages files and directories
- maintains disks
- configures hardware
- optimises the use of memory
- speeds up programs
- customises applications.

There are, usually, two ways you can work with operating systems – by using special menus (called shells or windows) or by typing commands at the command prompt. The following are the operating systems most commonly used on PCs.

5

- DOS (Disc Operating System) — one of the first PC based operating systems. It has almost disappeared as an independent system. Most of the current operating systems provide a command-line interface (known as a C:> prompt) to emulate DOS commands.
- UNIX — a multi-user, multitasking operating system which allows more than one user to work concurrently on several applications. Being originally developed for minicomputers, it is widely used today as a network operating system, especially in conjunction with the Internet. Many flavours of UNIX are available free, making it an instrumental part of open source movement. One of the best known examples is **Linux** (pronounced *lih-nucks*) a version of UNIX that runs on PCs. It was developed by Linus Torvalds (for whom it is named) along with numerous collaborators worldwide. Linux is distributed free, and its source code is open to modifications by anyone who chooses to work on it, although some companies distribute it as part of a commerical package with Linux compatible utilities.
- Windows — an operating system introduced by Microsoft Corporation in 1983 as an alternative to DOS. Windows is a multitasking graphical user interface environment. It has a self-contained operating system for personal computers (i.e. Windows ME), workstations (i.e. Windows 2000 Professional) and Network Servers. Windows provides a standard graphical interface based on drop-down menus, windowed regions on the screen, and a pointing device such as a mouse.
- OS/2 is an operating system from IBM, developed for Intel-based PCs only.

Computer Languages

All computer software is written using a programming language. Each language has its own instructions (vocabulary) and syntax, and therefore each has its pros and cons. Nevertheless, there are several levels (called generations to reflect the time-related evolution) of languages distinguished today.

The *first generation* are based around binary numbers 0 and 1.

The *second generation* are specific to a particular computer or micro-processor.

The *third generation* consists of many older and newer languages such as Basic, Cobol, FORTRAN, Prolog, LISP, C, Ada, etc.

The *fourth generation* languages are distinguished by a syntax using a more human-like vocabulary of commands. Examples are dBase, Lotus1-2-3 and similar software development tools.

The *fifth generation* languages are being developed by the Japanese as part of the new millennium project to build a fifth generation computer. The ambition is the use of artificial intelligence techniques to create a computer tool that will be able to solve a given problem once it has set out the data. The emphasis is also placed on the better human–computer interaction, with speech and images as main communication tools. The PROLOG computer language has been selected as the main development language.

Applications Software

The term applications software refers to programs that deliver a specific service, such as word processing, drafting and data management across several functions. For example, Lotus 1-2-3 was the first integrated software, combining three function services – spreadsheet, file management and graph generation facilities – in one program.

Systems Development

Managing an organisation, especially if it operates from and shares resources with different locations (i.e. flight ticket booking can be done on-line from virtually anywhere today), requires a special computer solution to take into account many specific operations, procedures and functions required to support distributed and speedy decision making. An example in construction would be software designed to deliver the paperless project. All partners use common software, e.g. humming bird.

Developing computer system solutions for an organisation is a difficult task, which involves a cycle of activities. The systems development life cycle (SDLC) is a kind of code of practice for those who are investigating how to improve procedures, tasks or organisation performance. The system development life cycle has five phases (Capron and Perron, 1993):

1   Preliminary investigation – determining the problem
2   Analysis – understanding the existing system
3   Design – planning the new system
4   Development – doing the work to create the new system
5   Implementation – converting to the new system.

Computer Communications

Broadly speaking, communication is the transfer of information between two or more parties. Effective communication implies that the receiving party both receives *and* understands the message. Little or no information can be exchanged between two people speaking different languages.

As far as computers are concerned, the physical separation of two devices may vary from a few metres to several thousand kilometres. The transmission

Source   Destination

		1		1
1		0		1
2		1		2
3		0		3
4				4
.				.
.				.
.		1		.
n				n

(i.e. computer ⟶ disk
disk ⟶ bus)

Source   Destination

(i.e. computer ⟶ plotter
computer ⟶ modem)

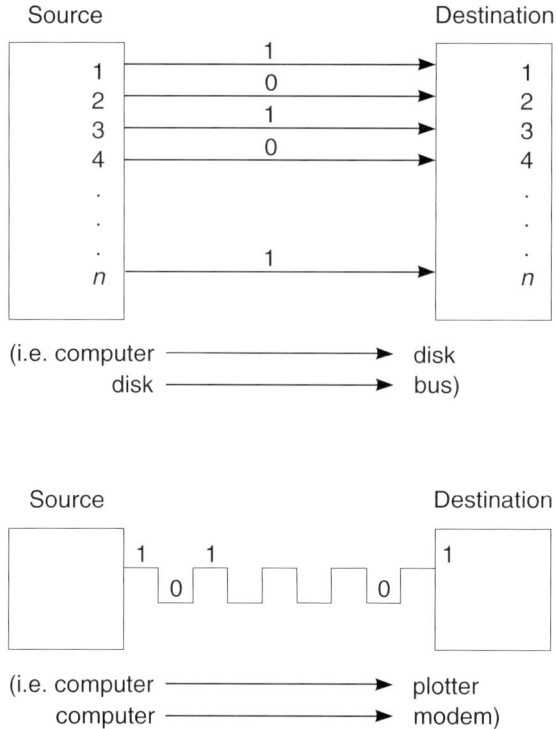

Fig. 1.2   Data transfer nodes: (a) parallel; (b) serial

medium may be a wire or a radio-based channel. The more general term 'data link' is used to describe the link connecting two parties.

Where machines are close together the information can be transferred by parallel transmission. When machines are far apart a serial model of transmissions is used along an individual wire. These are shown in Figs 1.2(a) and 1.2(b).

Over Long Distance

- It is impracticable, both physically and commercially, to lay a cable just for the purpose of linking two personal computers.
- There are technical difficulties in transferring serial data over long distances without boosting the signal at regular intervals.
- A separate cable would have to be laid for every destination with which a sender wished to communicate.

Consequently, it has become standard practice to use an existing network of cables – the telephone system – to transmit data. In order to connect the computer to a telephone line a modem is used.

The main problems lie in the fact that the telephone network was originally designed to carry audio signals – i.e. the human voice – so that digital data must first be converted (modulated) into analogue form for transmission and then converted back (demodulated) again at the receiver. This is the function of a modem (MOdulator–DEModulator).

A modem is basically a device (either internal card or external little box) that enables one computer to talk to another computer down a telephone line, moreover the two machines do not have to be the same.

ISDN

ISDN (Integrated Services Digital Network) technology was developed to speed up modern communications. It is a worldwide telecommunication network. Using ISDN allows you to receive and transfer very quickly large amounts of computerised information, such as drawings, images and video files.

ISDN technology requires special lines provided by telecommunication companies. Although cost of transmissions for such lines is similar to telephone lines, the speed of file transfer makes running costs much lower than modem-based communications. It means that despite higher initial 'connection investment' this service benefits companies with remote offices.

DSL (**D**igital **S**ubscriber **L**ine), a recently developed (late 1990s) digital communications technology that, similar to ISDN, can provide high-speed transmissions over standard copper telephone wiring.

Computer Networks

Computer networks are used to interconnect distributed computer users. When the computers are spread over a small area such as an office or building, the network used is known as a local area network. On the other hand, when the computers are distributed over a wider geographical area, such as a region or country, the network is know as a wide area network.

Communication developments have interconnected sets of wide area networks creating global area networks known as the Internet. There are three types of network;

Local Area Network

A local area network (or LAN) is a *dedicated* communication network for linking together computing equipment over a small geographic area such as an office building, factory or college.

The machines can be connected up in one of four configurations: bus, ring, star and hub (see Figure 1.3):

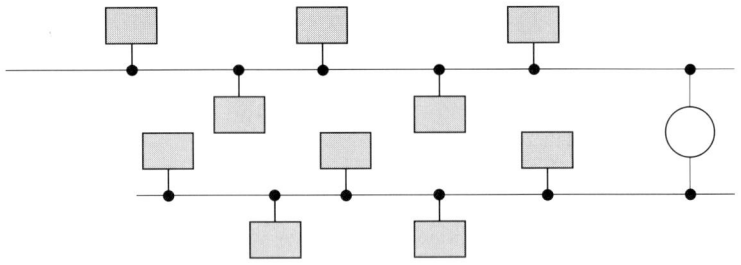

(a)

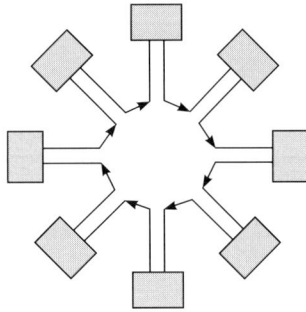

(b)

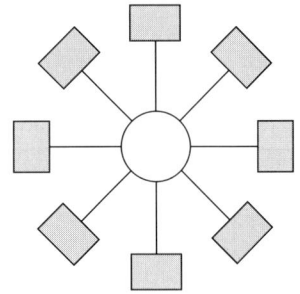

(c)

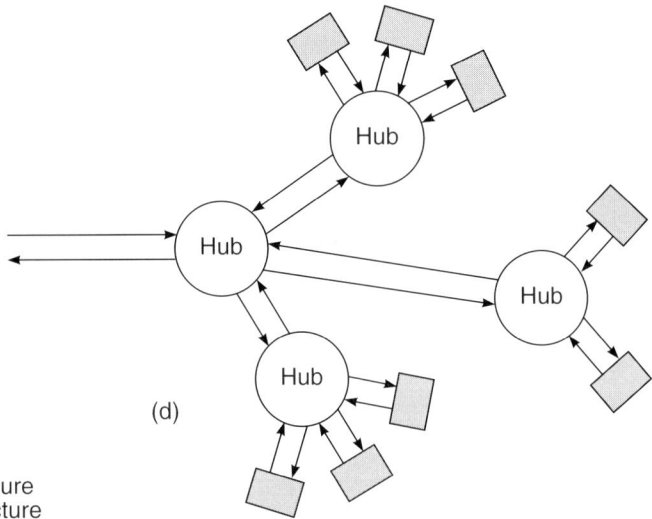

(d)

(a) Bus structure
(b) Ring structure
(c) Star structure
(d) Hub or tree structure

Fig. 1.3 Network Topology

A *bus structure* (Fig. 3.1a) can best be envisaged as a single line of cable along which computers broadcast their message.

A *ring structure* (Fig. 3.1b) is a connected cable forming a ring in which transmission takes place in one direction only around the ring, thus avoiding problems of collision.

A *star structure* (Fig. 3.1c) manages the LAN through a central or hub computer from which all the lines plus nodes radiate like the spokes on a wheel. A device wishing to communicate with another on the LAN will send its transmission into the hub computer from which it will be routed out to the recipient.

A *hub* or *tree structure* (Fig. 3.1d) is a variation of the bus and ring. Hubs can be connected in a hierarchical way to form a tree structure.

Wide Area Networks and the Internet

At the beginning, the telephone network was the only method available for transmitting data between user equipment located at different establishments. Data transfer rate is limited (to about 9600 bits per second), and, as with telephone calls, these are charged on a time and distance basis, making some transactions very expensive.

As a result, many large organisations established their own national and international private data networks, which typically used dedicated lines leased from the telephone authorities. Although networks of this type offer the user security, flexibility and ultimate control, they also involve high investment in purchasing or leasing the equipment. Such networks are therefore generally owned and managed by large organisations, such as major banks, who not only can afford the initial capital but also have large traffic of data to justify the investment. These networks are known as private enterprise-wide networks.

Following the establishment of public networks, facilities were developed enabling network users to communicate with each other. These networks, which link computers located in different organisations and/or sites, are known as wide area networks (WANs). The establishment and operation of a WAN requires agreed standards for access to and use of equipment from different manufacturers and, in turn, a set of internationally agreed standards to facilitate the interconnections. In the 1980's in the USA, the computer networks of a large number of universities and other research and government establishments were linked. The resulting internetwork has been later extended to incorporate internets developed by other agencies and countries. The combined internetwork is known as the Internet or 'information superhighway'.

The Internet is a network of sites or *de facto* the global area network (GAN). Each Internet's site (educational, military, commercial, etc.) is a LAN or a WAN network itself. So each e-mail has a suffix which defines its location, so .com (company), .ac (academic), .gov (government) and .org (an organisation which is not a company, or university, etc.). The e-mail is the most ubiquitous use of the Internet.

File transfer protocol

File transfer protocol (or FTP) is another method of transferring files from one computer to another. Although electronic mail makes excellent use of the network infrastructure, it mainly provides a one-to-one or one-to-many communication tool. Attempts to find a way of searching for, and access to, information available on networks resulted in the phenomenally successful World Wide Web. Vast databanks of information can be searched using special software interfaces, called search engines, available through all Internet browsers.

The Internet and the World Wide Web

At present the Internet (commonly known as 'the Net') has more than 100 million registered users with about 50 million people using it daily. The number of daily users is increasing at a rate of 1 million each month. The main users and information providers are from the USA and Europe.

Access

Very significant improvements in means of accessing Internet information are becoming available. The World Wide Web (WWW, or 'the Web') was set up at CERN (Centre Européen de Recherches Nucléaires) in Geneva to provide a means of access to information for the physics community worldwide. The use of the WWW has spread beyond the boundaries of physics and has become the major access system to the Internet. Examples of these access systems are Internet Explorer developed by Microsoft®, and Netscape's Navigator, created by a private company.

HTML

HTML stands for Hypertext Markup Language and it configures text on the WWW so that it may be read as English or other Language being used. Links between pages could also be introduced, so users would easy navigate ('surf') between various documents.

XML *stands for* e**X**tensible **M**arkup **L**anguage. It offers Web developers and designers greater flexibility in organising and presenting data and information than is possible with the older HTML systems.

VRML

Virtual Reality Modelling Language (VRML) is also a generic language designed to allow three-dimensional objects (or 3D 'virtual worlds') to be depicted. VRML files describe a 3D space in a standard text like format which is interpreted by browsers. As a result, if you have an appropriate 3D viewer (or VRML browser) you can view and interact simultaneously from different locations with the 3D model of design or virtual environment.

A VRML world is made up of simple shapes such as cubes, cones and spheres, grouped together to form more complex objects. If required these can be modelled using common package (such as AutoCAD 3D Studios etc).

Client–server concept

Central to use of the WWW is the client–server concept. Information is stored on a computer, which acts as a 'server', while access to the information is from a 'client' computer. The client is the PC or workstation in your office; the server is normally in a location remote from the client. You can put material onto the network and two of the main motivations for making information available on the Internet are:

- Information is provided for publicity purposes. It is a mega-advertising medium that is not under commercial control. The situation has probably gone too far to enable reversal of this trend and it seems very unlikely that the Internet will come under commercial control ever. With the cost of getting connected and putting information on the Internet being relatively cheap, many organisations have shown an interest in the Internet not only as a way to promote themselves, but also as a medium for commercial service provision.
- Organisations or people who put information on the network potentially get back more than they put in. The investment in effort to set the information up is more than repaid by what can be received via almost unlimited and almost instant access to other sites.

It is important to note that though the Internet is almost 'ungoverned' and 'unregulated' territory, all rules and regulations that exist in the user's country apply to Internet use (including copyrights!).

Intranets

Many companies that started using the Internet as a way of communicating information to and soliciting services from the outside world very soon realised the benefit of the technology for the sake of their own employees. For large and geographically spread organisations the e-mail and Web browser cannot only speed and facilitate internal communications but can also include drawings, pictures, sound, video and other available media creating the true 'paperless office'.

The main advantage of an intranet network (as opposed to a local area network) lies in the ability to communicate to any user connected to the Internet with almost no regard to the user's computer maker or software configuration.

The growth of intranets has increased considerably in recent times. It is growing so quickly that it is impossible to assess how many intranets are in use. An estimated 100,000 intranets were in use in 1995, with an expected 5 million to be in operation in 2000 (Hills, 1997). There are many ways in which intranets have been used. Most of the use is around transfer of information. Internal e-mail is most common as well as the electronic noticeboard. Company manuals, policies, newsletters, training information, job advertisements and reports are the most popular items put on a corporate intranet.

Ford Motor Company's Intranet

An interesting case of the introduction of the intranet approach to the Ford Motor Company is described by Stuart (1997) and Gurton (1998). For the company, which has over 370,000 employees and operates on a global base, the main drive to choose an intranet was to solve the problem of documentation and information distribution, access and management. (It was not unknown to take half a year to find a required paper-based document!) The intranet was launched in 1996 and quickly took off with 80,000 employees installing Web browsers to their computers. Ford set up training sessions, and most people learned how to use the browser within only ten to fifteen minutes.

Among the topics that can be found on the Ford intranet are news; employee information; the company's vision, mission, goals, products and processes; competitors; the company's phone books, training programmes and application forms; building layouts and maps; quality control procedures; photos from motor shows, etc. There are also simple procedures set up for placing internal orders and payments. Though there is no information available on financial return, Ford reports on considerable benefits achieved, including improved efficiency in accessing and retrieving information, reduced use of phone calls, vastly reduced paper consumption, quicker customisation of newcomers to projects, better communication to overseas projects running on 24 hours a day global basis. Among the problems which have arisen are language barriers with some overseas parties (particularly in eastern Europe); some workers being unwilling to learn new technology (mainly those approaching retirement), and others reluctant to use it because of, they claim, security concerns.

Cost of Setting up an Intranet

Keys (1998) has analysed the cost of setting up and running a corporate intranet by five large companies (in the gas supply, electronics and manufacturing industries). The three-year expenditure is about $1.5 million on average (it's higher for those who needed to invest in hardware for setting up the intranet, and significantly lower for those who had the infrastructure in place). However, all companies reported that the investment paid for itself in a fraction of a year. Though the starting point, organisational structure and following solutions were different, some indication can be derived from the following figures. Among main items for cost allocation are consulting (20–40%), content development and management (10–40%), personnel development and training (15–60%), software (3–72%) and hardware (3–30%).

A project-specific intranet, for a smaller company, can be set up for as little as $5000 (Mead, 1997). Despite this figure, there are very few intranets in use by construction companies. Stockyk (1997) reports on Drake & Skull, and Bovis in the UK, with three other examples described in the case studies in the appendices. Hannus (Hannus et al., 1996) describes an interesting case where, due to the tight schedule and long distance between the designers and the construction site, the intranet-based environment was set up to speed communication to the site and exchange design files: with overall positive experience, just the benefits of time and courier charges saved were more than enough to return initial investment.

Extranets

There is often a benefit in allowing external parties (such as clients) to access a company's intranet. Those intranets allowing selected external access are called extranets. The client accessing the extranet can see how the company progresses with a project and obtain required information, or provide a feedback to the company. An example of such use is the extranet of Federal Express Co. The company, specialising in swift package delivery across the world, allows its clients to follow their packages on the Web site (www.fedex.com) using a tracking number.

Intranet Security

There is always a risk that any computer network could be accessed by unauthorised persons from outside or misused by an employee. The internal misuse would not depend on the network type, but the personal motive to do so. Contrarily, it is easier to break into the Internet-based networks (either Intranet or Extranet) than the LAN, simply because the former can be accessed remotely. In practice, the software companies have developed quite efficient and reliable systems (called 'firewalls') to prevent the unwanted access of the network from the outside world (Harr and Siyan, 1996). With electronic signatures being developed, the identification of the addressee will be easier and even more reliable. However, it should always be remembered that there is no system (especially paper-based) which can guarantee you 100% security.

Benefits to a Construction Company

There are two levels on which a construction company can benefit from an Internet-based network. On a company level, it can improve general information provision for both internal personnel and external bodies. On the project level, it can allow sharing and exchange of project related information (for example, schedules, drawings, invoices, minutes of meetings, etc.) with parties involved.

References

AOUAD, G. and PRICE, A. D. F. (1994). 'Planning and IT in the UK and USA construction industries – a comparative study'. *Journal of Construction Management and Economics*, Vol. 12, No. 2, pp. 97–106.

CAPRON, H. L. and PERRON, T. D. (1993). *Computers and Information Systems: 'Tools for an Information Age'*. Redwood City, CA: The Benjamin/Cummings Co.

GURTON, A. (1998). 'Ford's IT heritage boosts intranet adoption'. Intranet Management Report, No. 9, p. 10.

HANNUS, M., HEKKONEN, A. and LAITINEN, J. (1996). 'Internet in construction projects and research'. In Z. Turk (Ed.), *Construction on the Information Highway, CIB W78 Proceedings, Publication 198*. University of Ljubljana, Slovenia, pp. 265–72.

HARR, C. and SIYAN, K. (1996). *Internet Firewalls and Network Security*, 2nd edition. Indianapolis: New Riders Publishing.

HILLS, M. (1997). *Intranet Business Strategies*. Chichester: John Wiley & Sons.

KEYS, F. (1998). 'The feasibility and benefits of the use of intranets to manage construction projects'. Unpublished MSc dissertation, University of Strathclyde, Glasgow.

MEAD, P. (1997). 'Project specific intranets for construction teams'. *Project Management Journal*, September, pp. 44–51.

PAULSON, B. C. (1995). *Computer Applications in Construction*. New York: McGraw-Hill.

STUART, A. (1997). 'Under the hood at Ford'. *Webmaster*, June.

STUKDYK, J. (1997). 'Internal affairs'. *Building*, May, pp. 44–7.

Chapter 2

Databases and Computer Graphics

Introduction

Data are, and always were, one of the most important assets of every organisation; catalogues and filing cabinets are an integral part of every construction firm. We store data and information mainly because we want or expect to *reuse* them. Therefore, we organise the storage facilities in a way that allows the *retrieval* of required data as quickly as possible. There is nothing more frustrating than looking for a particular drawing in an archive of an engineering company and discovering that the drawing is not where it is supposed to be! Going through all the stock to find the drawing is a mighty time waster.

It is, therefore, not surprising that as soon as computer technology provided data storage and access facilities (at the end of the 1950s/beginning of the 1960s), the file and data management systems were a focus of IT development.

Not only is storage and access to information an issue but how data is presented is vital to its acceptability. Here Computer Graphics are an essential part of a construction company's IT activities. Advances in computer hardware, especially in *raster* (TV-like) display technology, bring together two very related but, until recently, separate disciplines: computer graphics and image processing. Foley et al. (1996) state that while computer graphics deals with the 'pictorial *synthesis* of real or imaginary objects from their computer-based models', the image (or picture) processing concerns 'the *analysis* of scenes, or the reconstruction of the models of 2D or 3D objects from their pictures'. Pictorial synthesis tasks are often identified in construction and design examples of analysis and reconstruction involving image processing in construction are: remote sensing in surveying, sensing and computer vision in robotics, pattern recognition in geographical information systems (GIS) and photogrammetry.

As soon as computer graphics techniques allowed storing, manipulating and displaying objects' geometry, colour, related text and other properties, many software applications were developed to support construction activities. Computer-aided design and Drafting (CADD) is one such example. The success of CADD and other practical applications has triggered, and at the same time been influenced by, the ability of low-cost computer tools to support interactive engineering tasks involving large amounts of 2D and 3D data and information. This, combined with the dramatic increase in microprocessor power, has resulted in many effective tools to support data preparation and presentation being produced, such as digitizers, scanners, plotters, large and high-resolution display projectors and screens. These tools are now not only 'affordable' but also well supported by effective software tools. Some of these tools and their construction application, such as visualisation and multimedia videoconferencing, will be briefly reviewed in this chapter. Others, namely virtual reality and GIS, deserve more attention and will therefore be described in a separate chapter (see Chapter 3).

Database Fundamentals, Models and Structures

According to Elmasri and Navathe (1989), a database is a collection of related data, where data are known facts that can be recorded and that have implicit meaning. In order to be recorded and accessed, data are organised, or presented, in several levels (see Figure 2.1).

This structure is a function of the internal representation which, in turn, is dictated by the computer specifications and software. Understanding of internal representation is important for those who write computer languages but is less intent to the practitioner. As we can drive cars without understanding the engine, we can use software without understanding bits and bytes. But it can be useful to understand the principles of databases.

A *database* is a coherent collection of logically related records or files. It is designed, built and populated with data for a specific purpose and has an intended group of users and some applications in which these users are interested. A project database, for example, consolidates data from separate files such as schedule files, resource files, subcontractor files, cash-flow files and others. It aims to facilitate a project manager's task of delivering a project on time, within an allocated budget and according to the project requirements.

A database can be of any size and of varying complexity. For example, a project time schedule usually has 100 or fewer records, while a materials database (or catalogue) may contain several thousand records. A database of the DVLA to keep track of UK vehicle registrations will have about 50 million car records. Assuming each record consists of 100 characters and assuming an average of three owners per car life, we would get a database of

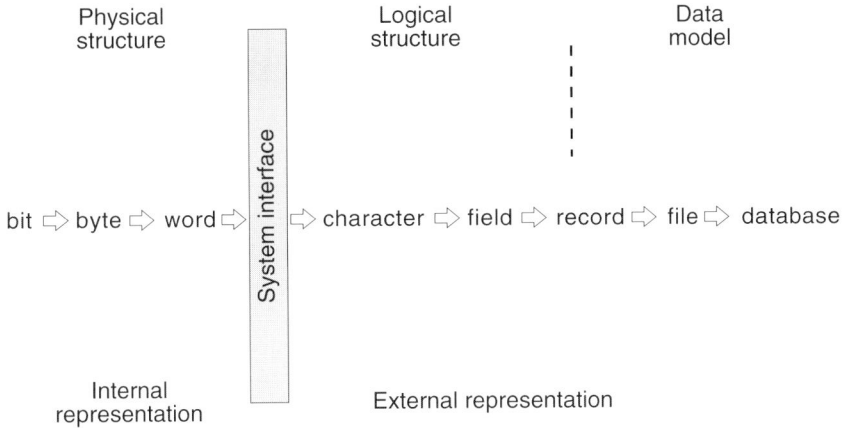

Fig. 2.1 Data representation levels

$3 \times 100 \times (50 \times 10^6) = 1.5 \times 10^{10}$. This huge amount of data must not only be organised and managed properly, but must also be available for search and retrieval quietly and on-line (for police patrols, for example).

A database can be generated and maintained manually or by computer. The telephone directory, materials catalogue or library catalogue are examples of manual databases. Duplication of data (for example, for code or name search), difficulties of update, and cost of consistency are the main problems of a manual database. These problems are alleviated in the computerised solution.

A computerised database is created and maintained usually by a group of programs called a database management system (DBMS). An example of such a system for cost estimation and tender preparation is presented in Appendix A.

Data Models

One of the fundamental characteristics of a database approach, as presented in Figure 2.2, is 'hiding' data storage details from users. Data types of, and relationships among, the records stored in a database are described by the logical data structure of a database called a data model. A specific data model is employed by a DBMS package to provide users with quick and easy access to the information stored in a database.

Therefore, the main criterion used to classify DBMSs is the data model employed by the DBMS. There are four fundamental categories of DBMS, which are based on hierarchical, network, relational and object-oriented models (see Figure 2.3).

19

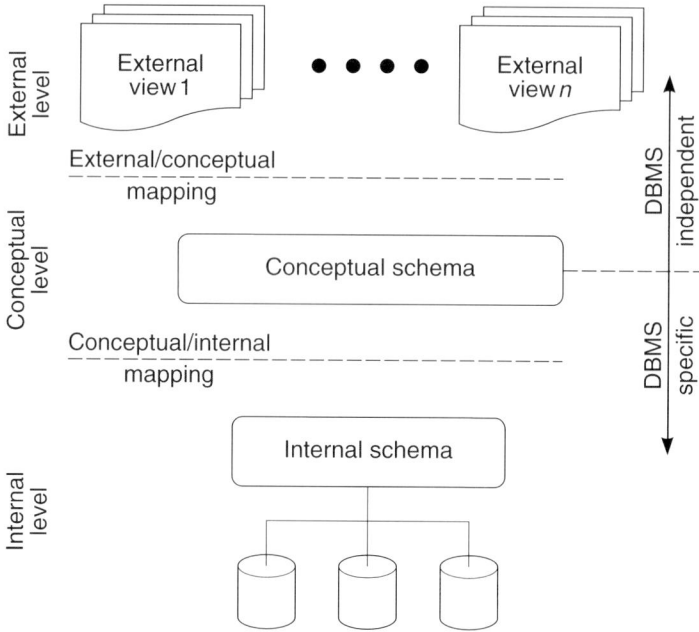

Fig. 2.2. Database approach

The *hierarchical* model represents data using hierarchical tree structures. Each hierarchy represents a number of related records, arranged in multi-level structures, consisting of one root record and any number of subordinate levels. This type of model is suitable for many structured organisations and business operations. It allows quick search and access to records.

The *network model* represents data as record types with more complex many-to-many relationships. This allows, unlike hierarchical models, flexibility in describing different structures and relationships.

The *relational model* simplifies data handling (storage, update, etc.) but also provides more flexibility in database and organisation maintenance and use. This model was developed in the 1970s and became the most popular model adopted by many DBMS packages (i.e. Oracle, dBase, Access).

The *object-oriented model* is a relatively new but very powerful high-level conceptual data model. This model allows you not only to handle more complex types of engineering data (e.g. graphics, geometry, pictures, etc.) but also to apply methods and processes on these data (i.e. design procedures). This is a result of an object-oriented programming approach that allows inheritance of attributes and encapsulation of data to be achieved (see more on object-oriented programming in Chapter 3).

(a)

(b)

(c)

(d)

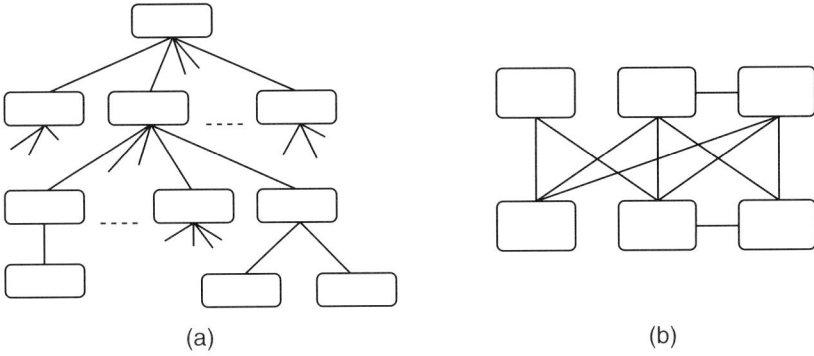

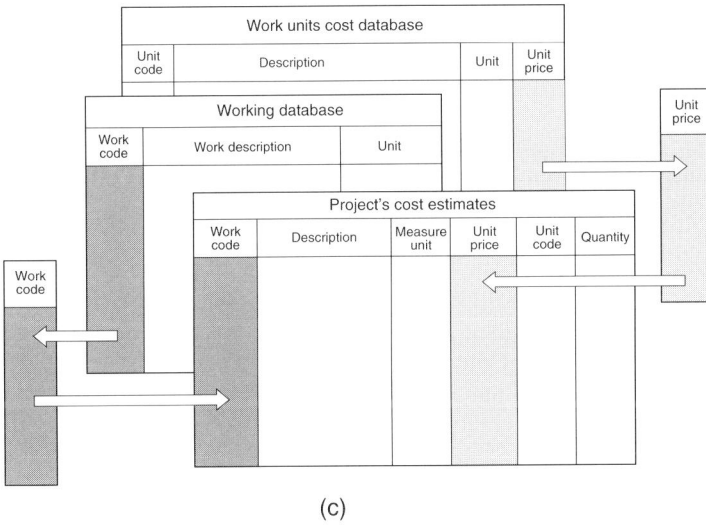

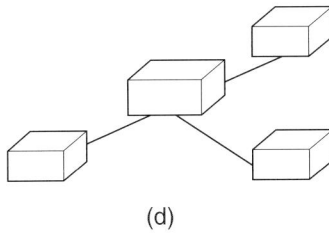

Fig. 2.3.   Data models: hierarchical; network; relational; object oriented

Database Types

There are several ways to classify databases. As far as overall application areas are concerned, databases can be presented as either engineering or managerial databases. These two types will be presented in more detail below. As far as the type of data is concerned, we can distinguish traditional, text, image and multimedia databases.

Traditional Databases

Traditional databases hold data as records and files (e.g. cost database, personnel database and similar). This was the first available type and today the majority of databases are of this type.

Text Databases

Text databases developed naturally from the traditional type when computer technology was easy and a cheaper way to create, distribute and store textual documents electronically. Examples of this type are bibliographic data, publications, documents, regulations, standards, etc. These are often available on-line through information providers or on CD-ROMs (Compact Disk, Read Only Memory). For example, Technical Indexes Ltd, jointly with the Royal Institute of British Architects, produces the Construction Information Service which includes full text databases covering most aspects of building and engineering design and construction.

Image Databases

Image databases are a new outgrowth of CD-ROM technology. These databases can contain not only text but also photographs, pictures and animated video sequences as digitised images. Many electronic encyclopaedias, picture galleries and museum expositions are available on CD-ROMs or even on-line.

In business areas, many existing documents (such as customer corres-pondence, orders, invoices, catalogues, manuals, drawing archives) can be optically scanned and stored as document images. Image database management software will allow employees to quickly search, retrieve and display documents from such databases.

Multimedia Databases

Multimedia databases are the most advanced and novel area in DBMS. They will allow not only images, but also 3D models and sound to be stored and manipulated.

Engineering Databases

An engineering database (often called a CAD or CAD/CAM database) is one of the components of computer-aided engineering design environments. Since CAD/CAM packages and network facilities have allowed interactive sharing of design information as well as data transfer, the engineering database became the most important resource of an engineering firm.

Chorafas and Legg (1988) say that the typical design-related information of such a database would include:

- a model of the design object
- text, data and graphical information elements
- properties of the object
- information describing the state of the design process
- design documentation.

A critical advantage of such databases is that construction design is a complex exploratory interactive process that is usually conducted by several professional teams. The design product (and its data) continuously evolves during such a process. As a result, construction design data are very different in comparison with a conventional database for the following reasons:

- construction design data are characterised by many different types of component
- design data records are long, with a variable length of textual information and compared to short, fixed-length classical databases
- temporal and spatial relationships are unique and important in design data representation
- design changes and updates are not predictable and are likely to affect many components or construction elements; these updates are very difficult to automate
- design management and co-ordination is complex and different from project to project
- there is a high level of duplication of design data, especially between architects, structural designers, HVAC designers, etc.

All this makes it difficult to capture and represent construction design data using conventional (as described above) data models. Though some CAD software packages attempt to solve part of the problem, the comprehensive solution has yet to be discovered. Will object-oriented technology do that? The answer to this question at the moment remains unknown.

Managerial Databases

A managerial database is a misleading name. In fact, this term encompasses several types of databases, many of which are an integral part of every (construction) organisation. These databases are part of information systems (IS) and decision support systems (DSS). The main types of such databases can be assigned to two groups and will now be reviewed briefly.

The Location Group

The location group consists of end user, external and distributed databases:

- *External databases* are usually privately owned on-line databases or databanks. They can be accessed by end users through commercial information providers or services (hosts) for a one-off fee or annual subscription.
- *End user databases* are those developed by the users for themselves. These databases are usually developed using commercial spreadsheets or DBMS packages and consist of text, contact names, prices, specs, etc.
- *Distributed databases* are those held at local offices, branches or sites of an organisation. These databases will usually be part of an organisational common database, although they can include site-specific data. Interaction, consistency of data and update timing are the main problems of these databases.

The Function Group

The function group consists of operational, management and information warehouse databases:

- *Operational databases* are often called production or subject area databases. These databases will contain data supporting the organisation operation, such as a personnel database, subcontractor database, project database, etc.
- *Management or information databases* will contain data and information needed for managerial decision making and will be a part of the decision support systems in executive information systems.
- *Information warehouse databases* are used to store historical data extracted from the operational and management databases of an

organisation. The data are organised in a standard way to be used by managers and users across an organisation for various activities.

Database Management

Advances in computing technology have caused an 'information explosion'. It is easy and not expensive to produce, copy and distribute data in electronic form. Managing this data in a manner that best contributes to business objectives has become a complex problem (Date, 1990). To overcome this problem a database management approach is a necessity. The main goal of such an approach is to reduce the duplication of data and increase their accessibility.

Users should be provided with inquiry and reporting capabilities to allow them easy access to the information they need. Investment will be required not only in hardware, software and storage facilities, but in database data administration and management, often called data resource management (Tom, 1991).

The latter will need to provide the security and integrity of an organisation's databases by reducing their vulnerability to errors, fraud and failures. O'Brien (1993) distinguishes three functions or components of database management; database administration, data administration and data planning.

Database Administration

This is a management function responsible for the proper operation and use of database technology, as well as technical support. The main responsibilities will include designing and maintaining databases as well as monitoring performance, enforcing standards and data definitions, and addressing security issues.

Data Administration

Data administration involves the collection, storage and dissemination of all types of data. It should be organised in such a way that data can be shared with and made readily available to all organisational users.

The data administration function (unlike operational and technical database administration functions) 'treats' data as a strategic resource of the organisation. It therefore focuses on how planning and control of data can support the organisational function and achieve strategic objectives.

An organisational data administrator (often called an IT manager) will, therefore, be responsible for data planning activities, developing policies and setting standards for database design, processing and security arrangements as well as for selecting database management software.

Data Planning

This is a major component of the strategic planning process in the organisation. This function focuses on data resource management and development within the organisational processes and plans, and on support of long-range planning commitments.

Integrated Databases

Two recent trends in computer technology have contributed positively to encourage construction organisations to use DBMSs. Increasing processing power of PCs and improved network infrastructure have boosted the development of distributed database systems. Such systems running in one location would use data from another database in a different location. For the users, the database they access will appear as a single integrated database (Elmasri and Navathe, 1989). This facility fits very well the fragmented construction industry. On the other hand, the integration problem has become even greater. First of all, in order to share data and integrate information they should be computerised!

The last two decades of computerisation resulted in a majority of organisations using both computerised and paper-based design and management working environments. There are two main reasons for avoiding full computerisation (Charafas and Legg, 1988):

- established processes are very difficult to change
- many historical data (e.g. documents, drawings) are on paper.

On the one hand, technology has advanced to solve part of the problem by producing scanners for documents and drawings (there are still many problems to be solved in 'intelligent' conversion of manual drawings into CAD readable files). On the other hand, the process of integration is 'an Achilles' heel' of the construction industry.

Before looking at a few possible ways forward, we'll look at the main reasons or benefits for an organisation to 'switch' to the integrated handling of data and information (Chorafas and Legg, 1988):

- cost reduction − this is likely to be achieved in the long rather than short term
- quality improvement − this is a direct result of computerisation
- co-ordination improvement − the full benefit is achieved when the users are networked
- construction and managerial productivity increase − 'Productivity and quality go together'.

Much of the current research concerned with integration issues also examines the construction processes and practices (i.e. project procurement routes from inception to completion). Many researchers believe that these should be changed (or re-engineered) and computerisation seems to give a good push to re-examine the way in which many organisations operate (Brandon and Betts, 1995). This is a complicated issue. Examples of a possible way forward are offered in the case studies given in the Appendices and in later chapters.

Computer Graphics Applications

The wide variety of computer graphics applications has resulted in a number of classifications used to categorise them. These, according to Foley et al. (1996), can be done by:

- type of object to be represented (2D, 3D), and the kind of picture to be produced (e.g. line drawing, colour image, shaded image, etc.)
- type of interaction, which determines the user's control over the object and its image (e.g. interactive design, off-line plotting, animation sequencing, etc.)
- role of the picture (is it an end product, a map or drawing, or a specific representation of the object being designed or analysed?)
- relationship between objects and their pictures (e.g. assembly, hierarchies, etc.).

Visualisation

'One picture is worth a thousand words' says an ancient proverb. This view is directly reflected in *presentation graphics*, which is primarily concerned with the communication of information that is already understood (e.g. Gantt charts, pie charts, graphs, etc.). *Scientific visualisation*, however, is concerned with exploring data and information graphically and is often referred to as visual data analysis (Earnshaw and Wiseman, 1992). Scientific visualisation systems are combinations of tools, techniques and technologies to enable insights into data and processes that are impossible by other means. A few examples, already available as commercial applications, are stress distribution in a structure, turbulence effects in fluid flow, and lighting or heating effects in a design space.

Tools and Techniques

Visualisation tools and techniques utilise aspects of computer and cognitive sciences, image and signal processing design. These are comprehensively described by Brodie et al. (1991), presenting visualisation as an important tool for industry and scientific research. As far as applications of visualisation are concerned, the following main cases can be distinguished: data visualisation, volume visualisation and process visualisation.

Data visualisation techniques and tools are important in cases with a large amount of data produced by external devices (e.g. remote sensing for GIS, automated recording of a structure behaviour during a test, etc.), or computer models (e.g. finite element analysis, flow simulation, etc.). The output in these cases is usually presented as a picture, map or graph with a different colour used to identify extreme areas, boundaries or specific values.

Volume visualisation techniques are used in cases where not only distribution of data on the surface is important, but also depth and location. Brain scanning is the best illustration of this category. Some construction examples are: non-destructive and non-invasive structural tests, soil investigation, 3D models generation, or reconstruction using photogrammetry.

Process visualisation cases are usually very complicated and will require special techniques such as animation and visual simulation.

- Animation is a recorded sequence of still objects or computer generated images that produce the same result every time it is presented.
- Simulation is the process of modelling the behaviour of a system and conducting experiments in order to better understand the system or predict the output of its future operation.

Computer Animation

According to Sleurink (1995), the two main methods of computer animation are frame-by-frame and real-time animations.

Frame-by-frame

The frame-by-frame method is based on individual preparation and separate storage (on disk or video tape) of each animation frame before viewing at 25 frames a second (30 frames a second in the USA). Being able to dedicate whatever time is needed to produce one frame, this technique allows you to produce very realistic ('photo quality') and/or special effect sequences.

Real-time animation

In real-time animation, the computer generates and directly displays images at the viewing speed. This technique is used where interaction is required. Flight and drive simulators as well as architectural 'walk-throughs' are the most popular implementations of this technique. A powerful computer with a large memory and a special graphics card will be required in order to make such animation realistic in a 3D environment.

Simulation

Simulation can also be categorised into two main types: *discrete simulation*, which mainly deals with queuing systems (such as product-flow through a checkout), and *continuous simulation*, which models systems that change continuously through time (such as flight or building construction simulation).

Hybrid simulation software, which copes with both sets of problems, is also available (Sleurink, 1995). Being one of the most powerful problem-solving tools, simulation is used widely in many engineering areas.

A variety of visualisation tools are available today to support design, traffic control, building performance evaluation, emergency situation assessments, training, etc. Not only is the visual simulation a very cost-effective training tool (such as flight simulation) but it can also provide solutions to problems that are otherwise impossible to achieve. For example, simulation of the construction process during design stages can help in identifying buildability problems, and save resources. The unique ability of computers to produce real-time interactive simulation is fully exploited in virtual reality systems, which will be described in the following chapter.

Video Conferencing

The term 'video conferencing' is not a new one. Traditional video conferencing has been available since the appearance of TV and communications technology. It requires expensive delivery and reception installations and entails high transmission costs.

One of the existing examples of such a large-scale system is satellite-based 'interactive television', i.e. one-way video, two-way audio. This allows for broadcast from a central point to many different locations regardless of distance (Coventry, 1995). Small-scale, or PC-based, systems are now available and are growing worldwide since the introduction of video compression technology (MacLeod, 1996).

Video conferencing can be defined as telecommunications and video technology allowing individuals and/or groups in remote locations to participate in meetings in the form of a live video as well as share data and applications. It is often a successful and definitely cost-effective alternative to traditional meetings.

Cunningham (1996) presents three ways of video conferencing:

- *One-to-one* meetings, involving full two-way audio and video communications for both sites (known as point-to-point communications).
- *One-to-many* communications involving full audio and video broadcast from a main site, while other sites are only able to send audio or text. Examples of this are interactive TV and remote lecturing.

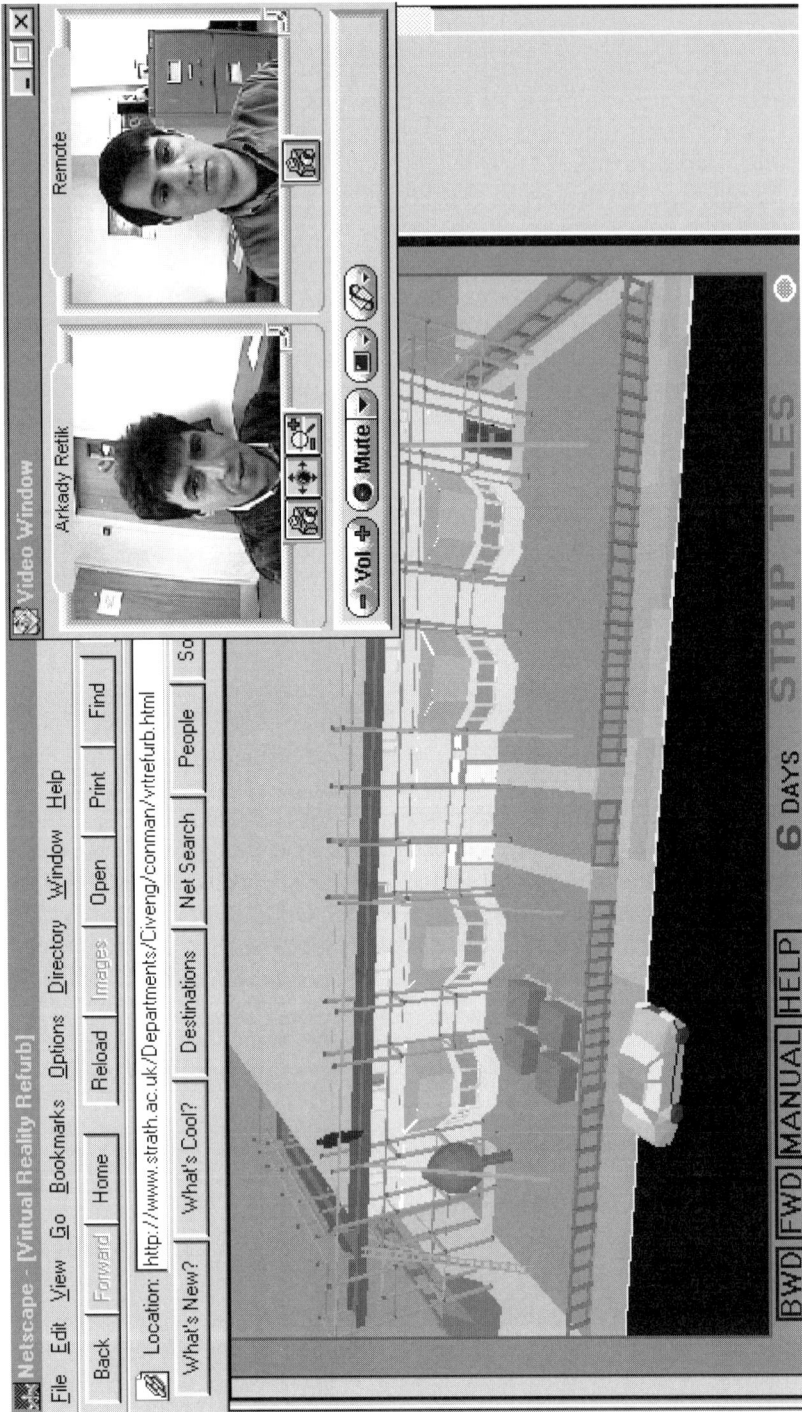

Fig. 2.4.   Typical PC-based desktop videoconferencing system

- *Many-to-many* conferencing, providing video and audio facilities between more than two sites (also known as multi-point communication). Most of these systems allow only one site in a conference to be seen 'live' at a time, though some (called MCUs – multi-conferencing units) will let multiple sites take part in a conference simultaneously (MacLeod, 1996).

Video Conferencing Equipment

A typical PC-based desktop video conferencing system is shown in Figure 2.4. The following hardware components are basic for every system:

- *network card* for connecting to the LAN or the telephone line
- *video card* for converting the images from the camera to digital form
- *camera* to capture pictures or scan documents. It can be tilted, zoomed and moved around as well as controlled by one or more nodes
- *microphone* and *headphones* (or speaker) to provide audio facilities.

References

BRANDON, P. and BETTS, M. (Eds) (1995). *Integrated Construction Information.* London: Spon.

BRODIE, K. W., CARPENTER, L. A., EARNSHAW, R. A., GALLOP, J. R., HUBBOLD, R. J., MUMFORD, A. M., OSLAND, C. D. and QUARENDON, P. (Eds) (1991). *Scientific Visualisation – Techniques and Applications.* Berlin: Springer-Verlag.

BUTTERS, L., CLARKE, A., HENSON, T. and POMFRETT, S. (1994). 'The do's and dont's of video conferencing in higher education', UK AGOCG Report, HUSAT Research Institute, Loughborough University,

CHORAFAS, D. and LEGG, S. (1988). *The Engineering Database.* London: Butterworth.

CLARK, S., MAHONEY, G. and STRIVENER, S. (1995). 'A study into video conferencing using the Apple Macintosh platform'. The Design Research Centre, University of Derby, July.

COVENTRY L. (1995). 'Video conferencing and learning in higher education', UK Advisory Group on Computer Graphics (AGOCG) Report. Institute for Computer Based Learning, Heriot-Watt University, Edinburgh.

CUNNINGHAM, S. (1996). 'Briefing of video conferencing', Advisory Group on Computer Graphics, Report. Manchester Computing, Manchester.

DATE, C. J. (1990). *An Introduction to Data Base Systems,* 5th edition. Reading, MA: Addison-Wesley.

ELMASRI, R. and NAVATHE, S. (1989). *Fundamentals of Database Systems.* Redwood City, CA: Addison-Wesley.

FOLEY, J., VAN DAM, A., FEINER, S. and HUGHES, J. (1996). *Computer Graphics: Principles and Practice,* 2nd edition. Reading, MA: Addison-Wesley

MACLEOD, M. (1996). 'A room with a view'. *Communications International,* Vol. 23, No. 11, pp. 72–6.

O'BRIEN, A. (1993). *Management Information Systems: A Managerial and User Perspective*, 2nd edition. Burr Ridge, IL: Irwin.

SLEURINK, H. (1995). *The Multimedia Dictionary*. New York: Academic Press.

TOM, P. L. (1991). *Managing Information as a Corporate Resource*, 2nd edition. New York: HarperCollins.

Chapter 3

Technical Applications of Computer Software

Introduction

Having grasped the technical framework of operating systems and generic applications of information technology, this chapter discusses some of the applications of software which may be used to aid construction design, planning and on-site operations. In particular three specific applications are highlighted they are:

- Virtual Reality
- Geographic Information Systems
- Artificial Intelligence.

Virtual Reality Systems

Like the telephone, virtual reality (VR) technology is a communications medium, but it is also a tool for looking at information. It is one of the advanced computer graphics technologies dealing with visualisation. VR gives users an efficient and effortless flow of data, details and information in the most natural format possible – vision, sound and sensations are presented as an environment, part of the natural media of human experience and thought.

The term 'virtual reality' was introduced by Jaron Lanier, founder of VPL Research, in the USA, although Myron Krueger used phrases such as 'artificial reality' and William Gibson coined the term 'cyberspace' in his 1984 science fiction novel, *Neuromancer* (Machover and Tice, 1994).

The use of object-oriented techniques for creating virtual environments was a key for the breakthrough in credibility and applicability of this technology (Larijani, 1994).

Definition of VR

Various definitions of VR are suggested today: one we find both descriptive and short is by Pimentel and Teixeira (1993), which defines VR as 'the place where humans and computers make contact'. Similarly, Larijani (1994) defines VR as the convergence of computer simulation and visualisation that attempts to eliminate separation between a user and a machine. So, from these authors' viewpoints, VR is an interface between humans and computers.

The Human–Computer Interface

Historically, the computer interface has been designed to lever human capabilities up to the advantages of computers, not humans. All too often, it would seem that the interface has been a way for the programmer to control the behaviour of the user, rather than as a way for the user to control the behaviour of the computer. Indeed, if we need to communicate a message to another person we will speak to him or her. So, why do we need a keyboard to speak to computers? This is, in fact, an area where research and development efforts have already brought some solutions. You can today dictate your letters to a computer's word processor, or even give commands to AutoCAD software (e.g. 'draw circle', 'diameter 5', 'layer red', etc.).

However, as far as engineering is concerned, another restriction of computer technology is a flat computer screen. Working with such a screen is not the most convenient way for us. We prefer horizontal surfaces to work on. If we want to look at something, we prefer to have several different points of view. Just watch how your colleagues behave while they are 'staring' at computer screens from about 50 cm and moving around a 'mouse'! Such a screen barrier often limits our creativity and imagination. Though it's a glass barrier, it is very difficult to break it.

Now, with virtual reality technology, it is possible to create a complete environment that interactively responds to and is controlled by the behaviour of the user. The 'reality' of such an environment does not depend on whether the created virtual world is as real as the physical one, but whether the created world is *real enough* for *you* to suspend your disbelief for a period of time. Similar to being absorbed in reading a good novel, you stop considering the quality of the interface media and accept the computer-generated world as a viable one (just as you might accept the voice on a phone as real, even with a 'noisy' connection). This is called immersion into a virtual environment (VE). Such an effect can be achieved by either immersive or non-immersive VR systems.

Immersive VR Environments

To generate immersive virtual reality systems or environments, there are three basic elements required: immersion, navigation and manipulation.

To be immersed in a virtual reality system is to feel that you are experiencing an alternative reality from the *inside*, being a part of the virtual world. Immersion is primarily a function of hardware: many systems use a head-mounted display (HMD) to stimulate the visual sense. By presenting a slightly different image to each of your eyes, the HMD uses the phenomenon of binocular parallax to create 3D effects. The stereo headphones are usually added to give you cues about the source and direction of the sound (see Figure 3.1). Additional input devices such as gloves act like the HMD providing data on hand movement to the computer.

If immersion tricks you into thinking you are in an alternative reality, the navigation gives you the opportunity to explore it. Navigation is the ability to move about in and *interact* with a computer-generated cyberspace. This ability is very important for engineering applications.

The manipulation function gives the user of a virtual reality system the most important (from an engineering point of view) ability to manipulate the virtual environment. This may be very simple, i.e. just to open the door, turn on/off the light, push a chair, and so on. The objects should respond as though they were real objects and behave according to forces of gravity, friction, and so on. However, the major difficulty is to provide a user with

Fig. 3.1. Basic PC-based immersive VR environment

tactile and force feedback which lets you know if the object is hard or soft, as well as describe its shape, temperature and other characteristics.

Non-immersive VR Environments

A virtual world can also be explored without immersion. *Desktop* (see example in Figure 3.1 without HMD) and *Projection* VR systems (see Figure 3.2) retain the navigation and manipulation features, giving a user the ability to move around the virtual world and manipulate its components using a spaceball or simple joystick. Nevertheless, such non-immersive applications can have advantages over 'real' VR systems, especially in big virtual worlds like cities or construction sites.

Augmented Reality

Augmented reality (AR) is one of the important and rapidly developing areas of virtual environment. AR is a combination of the real and synthetic images (usually the real world and virtual objects). Such a combination can be viewed as a human–computer interface that augments the user's view of real-world scenes by superimposing or blending virtual objects (computer-generated images) with live video input, thus creating the illusion that the virtual objects and the real world co-exist in real time and space. AR technology has already been explored in areas such as medical research, military applications, entertainment, maintenance and engineering. The potential of AR applications to the construction industry seems to be extensive in areas such as design and construction control, planning and supervision, safety training, and maintenance.

AR, or hybrid VR, is a variation of virtual environments (Vince, 1995). As such, it utilises techniques that do 'not oppose the real and virtual worlds, but fuses them in an intimate symbiosis' (Imagina, 95). AR presents a virtual world, which enriches rather than replaces, the real world. It can, therefore, be described as the 'middle ground' between VE (completely synthetic) and telepresence (completely real) (Milgram and Kishinio, 1994). AR is therefore usually associated with enhancing an existing environment by adding objects to the real environment (Azuma, 1997) – see examples in Figure 3.3 (O'Connor 1998).

Milgram and Kishinio (1994) categorised all possible cases into a continuum referred to as the reality–virtuality (RV) continuum (see Figure 3.4). The real environment appears to the left of the continuum, representing any environment that totally consists of physical objects. The virtual environment appears to the right of the continuum, representing any environment that totally consists of virtual objects. AR lies close to the real world augmented by computer-generated data, and augmented virtuality

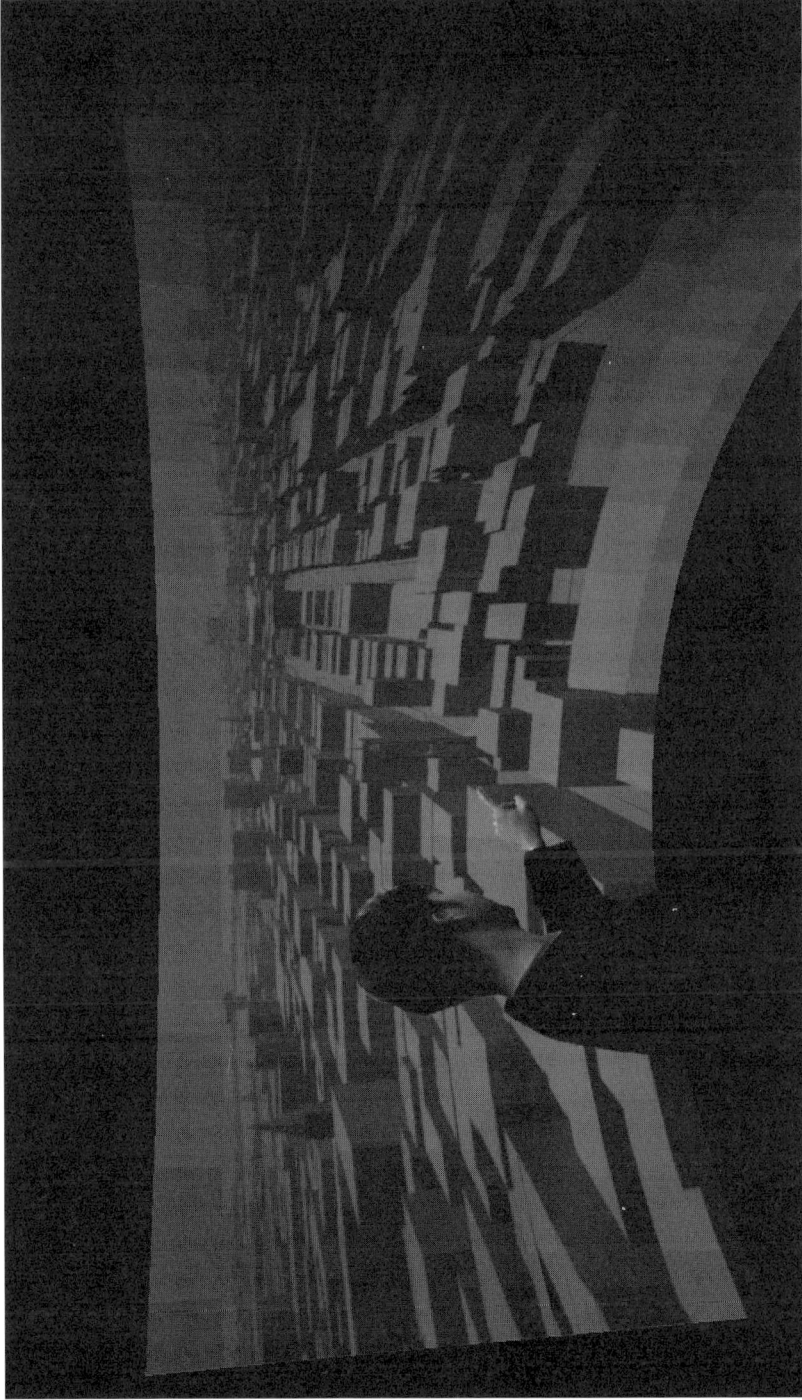

Fig. 3.2   Projection VR system

(a) Virtual Scene
Combining Virtual & Real Scenes

(b) Real Scene

(c) Augmented Scene

(a) Virtual Scene
Overlaying Virtual & Real Scenes

(b) Real Scene

(c) Augmented Scene

(a) Virtual Scene
Graphical overlap in a Real World Scene

(b) Real Scene

(c) Augmented Scene

Fig. 3.3   Augmenting Real World

(AV) lies close to the virtual world augmented with 'reality' obtained from, for example, video or texture mapping. Mixed reality lies in the middle region of the continuum and incorporates AR and AV. Milgram (Milgram et al., 1994) defines a generic mixed reality (MR) environment as one in which real-world and virtual-world objects are presented together within a single display that is anywhere between the extremes of the RV continuum.

Augmented Reality Systems

There are two ways to accomplish augmentation in a practical system: one is based on optical technology, another on video technology. The optical technology-based systems employ the optical see-through HMD, which allows the user to view the real world with the virtual world optically in front

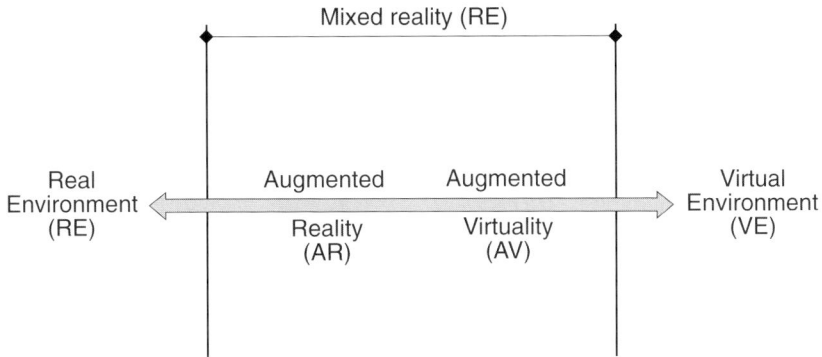

Fig. 3.4   RV continuum

of the user. This approach is similar to head-up displays currently used in military aircraft; however, optical merging is conducted in optical combiners of HMD as opposed to the 'window' type display in aircraft.

The video technology-based systems use video cameras to 'bring' a real-world view into a video composer in which the real and virtual-world images are merged. The resulting augmented images can be related either to the monitors in front of the user's eyes (closed view HMD systems), or to the external monitor display (called 'fish tank' systems). In the latter case, the tracking position of the user head and stereo viewing options would require additional special devices. Therefore, the fish tank systems are more suitable for static view cameras.

Both video and optical technologies are being extensively researched in the pursuit of the most efficient and effective AR system. Each method of achieving augmentation has its own particular advantages and disadvantages, which are discussed in detail by Azuma (1997) and (Holloway, 1997).

AR Application Domains

There are several research groups exploring AR for a variety of applications. The current domains for AR applications are computer-aided surgery, mechanical repair and maintenance, manufacturing, architecture, engineering, robotics, telepresence and military training.

In construction, Feiner et al. (1993) have developed a test-bed AR system for space frame construction. The system is designed to guide construction workers throughout the assembly of a space frame structure, ensuring that all members are securely placed in the correct position. Retik et al. (1997) are exploring the use of AR for planning and monitoring construction projects. The particular focus of this project is on remote access of a construction site using mobile telecommunications technology (see more in Chapter 5). There are other research works by Garrett and Smailagic (1998), Stalker and Smith

(1998), who are investigating the use of AR technology for monitoring bridges and structures.

It seems that combining AR technology with existing advanced computing and communications systems is capable of producing more visually powerful results for the end user, thus reinforcing integration, communication, visualisation, co-operation and management among project participants. It has yet to be determined whether AR is a truly cost-effective method of enhancing existing technology in its proposed applications, for the construction industry or for any other industry. An important factor in the implementation of AR systems into the construction industry is the provision of low-cost, efficient software and hardware.

VR Applications in Construction

Virtual reality as a graphics application can be used for a wide variety of uses (see, for example, Kalawsky, 1993; Pimentel and Teixeira, 1993). The training that can be accrued from VR, because of the real-time aspect, is the most realistic that the person could encounter without actually taking a physical part in the task that is being simulated. Moreover, a virtual project could be given to each contractor on site detailing what is required of them and how it is to be done. Contractors would be able to take a great deal of information from the simulation as well as being provided with a tool with which to train their staff.

In the creation of the project as a virtual project the manager is presented with an opportunity to test and operate the systems designed within the schedule. Depending on the time available, the value engineering and buildability aspects can be taken further than has previously been possible, leading to increased economies and better construction methods.

VR can be considered to be most beneficial wherever behaviour aspects of the construction site require simulation or where an added degree of visualisation is required. For instance, in Ribarsky et al. (1994), a virtual construction site is seen with the machinery on it made up of solid primitives. These primitives can move around the site environment interacting with the environment on user request to view effects. In a 3D model this could not happen, as there are no behavioural characteristics built into them. At the time of writing, VR has not reached the stage of development where it can be used to simulate the full complexity of the site environment.

VR can be used in construction for a wide variety of uses. Similar to the architectural applications, the immersive virtual reality systems provide the only way to learn and experience a design to be constructed. Then, during the planning and scheduling of the virtual project, a construction manager is presented with an opportunity to test different construction methods, applying the value engineering techniques and checking buildability aspects of the construction process.

Another important use of an immersive VR system would lie in its ability to provide training facilities for construction staff (DeOliveira, 1998) or even robots (Navon and Retik, 1997). Again, the training that can be accrued from VR, because of the real-time aspect, is the most realistic that the person could encounter without actually taking a physical part in the task being simulated.

Visualisation and simulation of the construction process using non-immersive projection VR may assist a planner or user in the better perception of a project as well as in the integration of other involved parties in the planning process (Retik, 1995). In large-scale projects, not only can the construction process itself be monitored, but also all auxiliary activities and on-site plant and equipment (Retik and Shapira, 1999). In addition the different locations of construction equipment and temporary facilities can be checked *in space* and can also be traced *in time*. In such cases, especially in projects where heavy plant is used, delays and interference may be prevented.

The planner can not only observe the simulated project, when performed with different scenarios, or at various stages of its execution, but also interact with a virtual project by walking through, moving equipment, etc. The ability to show a real picture of the work progress is very valuable for senior managerial staff, especially those dealing with the co-ordination of and resource allocation for several projects concurrently.

A VR projection system could be utilised further as a communication tool to keep members of the project team up to date with progress on site. The visualisation aspect means that they would all be as well informed as if they were visiting the actual project. Updates of the project could be sent via a computer network. Virtual reality could therefore offer an opportunity for the site to be run remotely from a head office. Using such a link, information from the site could be sent to the manager with an update of the virtual environment contained within it. This also allows the manager based in the on-site office to minimise the number of visitors requiring to visit the site for viewing purposes only.

Comparing the immersive or non-immersive types of VR a distinction can be drawn between their users within construction applications. Immersive is the better solution for training purposes, as it provides a far clearer and more exact representation of the real-site environment. Non-immersive would be ideal for simulating site operations, as it would allow activities and equipment within the site to be modelled.

Geographic Information Systems

Geographic information systems (GIS) are generally defined as a special group of decision support systems that integrate geographic databases with computer graphics. These systems are usually used to construct and display maps and related data in areas where resources and activities are related or assigned to geographic areas, such as urban management and control, roads and railway maintenance, electric power supply, and so on. Until recently, the cost of collection and management of geographic information prevented its wide use in civil engineering. However, the recent rapid development in information technology, especially in the fields of telecommunications and satellite communications, has opened up the ways to collect, manage and analyse more information with higher quality and at lower cost.

Oman (1996) describes the benefits of GIS in the planning and implementation of the Channel Tunnel rail link, and Thomas (1996) and Ray (1996) present the advantages of GIS for water companies. Bearing in mind that many operations in civil engineering are dependent on the analysis of spatial data (information traditionally recorded on paper maps), Parker (1996) argues that GIS should be utilised in civil engineering in a much wider way. GIS can be used not just for viewing a collection of maps, but also for integrating data and deriving qualitative information. These systems can improve planning, design, implementation and maintenance and management tasks as well as facilitate integration of many spatial and distributed civil engineering systems and operations (Mingus, 1996).

Geographic Information

Many civil engineering activities are dependent on geographic information. Hydrology, soil mechanics, infrastructure, transportation, construction and others will require some form of spatial (usually in the form of coordinates and maps) information. This information is often related to, and can be used and presented with, other sources and types of information, such as water catchment zones, density of population, street names, existing infrastructures and super structures, etc. In order to allow a systematic approach and secure future benefits from the integration and exchange of information, data acquisition, modelling and representation should be considered when GIS is set up in an organisation.

Data Acquisition and Representation

There are two types of data that can be distinguished: *background* and *company specific*.

Background data are:

- maps
- satellite images
- addresses, post/zip codes and streets
- rivers, canals, etc.
- geographic boundaries (local authorities, boroughs, counties, etc.).

These data are commercially available in many countries. The background data can be used as a basis for organising and storing the company-specific data, which need to be developed over time by a company or organisation.

The specific data may include:

- infrastructure (underground pipes, cables, etc.)
- substructure (offices, facilities, storage areas, etc.)
- geological data (bore holes, soil properties, etc.)
- network data (roads, communication routes, electricity lines, etc.)
- estates (land ownership, hazardous areas, etc.).

Either of the data types can be acquired and stored in two formats: *vector* and *raster* (Thomas, 1996). Vector format means that data can be stored as a series of coordinates and can therefore easily be analysed and modelled, allowing the rapid calculation of position, direction, length, and so on of any item. Raster format is usually a scanned or digital photographic image, which is much more difficult to analyse or to measure from. Raster images are usually used for displaying background maps as pictures.

It's often the case that background data are readily available in electronic format (either vector or raster). However, much of the company-specific data usually exist on paper maps. These need to be converted to digital formats manually (so called 'digitising') or by using raster scanning and vector conversion. Both methods are widely used and they are discussed in detail by Thomas (1996).

Data Modelling and Maintenance

Once GIS is set up, advantage can be taken of existing software for such things as ground modelling and route location, or a specific program can be created in-house. For example, Roorda (1996) Burns et al. (1999) presents the case of spatial information management for policy decision making in public infrastructure. She and Aouad (1996) describe the development of GIS for bridges monitoring and maintenance that integrates drawings and pictures as well as databases of bridges. A group of researchers from Stanford University (Basoz et al., 1999) describes the methodology for evaluating the effect of earthquakes on transport systems by utilising GIS and network analysis.

Whatever the use of GIS is, the data have to be maintained on a regular basis. Today, regular upgrades of background data are available from data providers, such as the Ordnance Survey or the Automobile Association (in the UK). On the other hand, the procedures for the company data maintenance and upgrade should be established and adhered to by the company staff.

Global Positioning System

A company-specific data collection is crucial for GIS. Until recently a land survey has been the main source. However, current availability of powerful and versatile satellite systems has brought to civil engineers a new technology: the Global Positioning System (GPS). Higher precision, compactness, ease of use and affordable cost allow GPS to be more frequently used not only for land survey and setting out, but also for data collection.

GPS is a three-dimensional measurement system for location or positioning points using data collected from orbiting satellites. The method is based on receiving radio signals emitted by the system of satellites. Processing the signals allows the determination of three-dimensional coordinates of receiver or differentiate positioning from another receiver. The US Department of Defense 24 Navistar satellites system (described by Simmons, 1996) enables accuracy within a few centimetres. This creates a great potential for GPS applications in highways profiling, land and hydrologic surveying, mapping and other civil engineering tasks providing data for a company GIS. Moreover, the recent introduction of real-time kinematics GPS surveying (Simmons, 1996) enables receiver coordinates to be computed in real time. This makes GPS a practical tool on a project level as well. Operation of the earth-moving plant, setting out, positioning of piles and boreholes, etc. are examples of the tasks for which this technology is already being implemented.

One of the strong points of the GPS technique is that it does not require inter-visibility between points. However, in order to receive signals from satellites, sky visibility is a necessity.

Satellite Remote Sensing

GPS has already successfully been implemented in civil engineering, and other possible ways to use satellite technology are under investigation, such as the practical application of satellite remote sensing as a source of information for civil engineering projects and GIS (Vincent et al., 1996; Scott, 1994).

Satellite remote sensing, or earth observation, was originally developed and used for both military and research purposes. Initially, the images were taken by optical sensors (cameras) and transmitted to earth where they were analysed and/or scanned. Recent advances allow the microwave technology to be used for scanning the earth surface, freeing remote sensing from the mercy of weather (clear sky) and light conditions. Vincent et al. (1996) predict that this development, being 'weather independent' and available in electronic form, will expose the satellite remote sensing technology as a valuable source of data for civil engineering in general and GIS in particular. Though, the authors warn, experienced and skilful interpreters are needed to make effective use of the source of data. Image recognition and pattern matching techniques could be used to automate the data acquisition in the future.

Virtual GIS as a Management Tool

Most of the information any organisation uses – 80–90%, according to Mingus (1996) and Parker (1996) – has some spatial content. Such information, therefore, can be related to, and originated according with, its physical location, such as an address, post or zip code, grid reference, street map, etc. Thus, a spatial index can provide a unique mechanism for the integration of distributed databases located in different organisations and/or supplied by various data providers. Moreover, Mingus (1996) claims that availability of powerful and inexpensive PCs coupled with 'an intelligent context-sensitive search facility to browse related available data sets, both geographic and multimedia' can and will create open GIS applications resulting in a 'virtual GIS'. A virtual GIS will require a standardised interface (similar to, for example, Netscape used for the Internet) to allow access to data in distributed geographic databases, located in different organisations around the world.

Three-dimensional visualisation can be used as a viewing system. This will not only enhance the decision support possibility but also increase the utility and integrative power of the system for civil engineering.

Progress in open GIS, reported by Cargil (1995) and Hering (1995), and standardisation efforts in this area, described by Mingus (1996), are significant and promising.

Artificial Intelligence And AI Related Approaches

The increased power of personal computers created a fruitful environment for advanced IT applications. It is not surprising, therefore, that much of the research and development has concentrated on looking for powerful computer tools capable of tackling many engineering challenges. This chapter describes some of these tools, in particular knowledge-based systems, the object-orientated approach and neural network systems, which especially relevant to construction applications.

Artificial Intelligence Applications

Mainstream computer applications have greatly outperformed humans in the areas of accurate and high-speed numerical input, processing, organising and retrieval of vast amounts of information. However, this is not the case when expertise and experience are required for finding solutions, especially in planning and design (Paulson, 1995; Coyne et al., 1990).

Artificial intelligence (or AI) is one of the branches of computer science that attempts to replicate non-numerical reasoning processes, which have been, until recently, considered inherent only to humans. Artificial intelligence does not attempt to enact human thinking processes, but utilises computer tools to approximate human (expert) behaviour. Rich (1986) defines AI as 'the study of how to make computers do things at which, at the moment, people are better'. Today AI-based systems are used in a number of areas, such as diagnosis, speech processing, natural language processing, games, image processing and robotics. However, AI technology has a notable significance for the areas employing complex heuristic (experience-based) knowledge, such as civil engineering (Dym and Levitt, 1991; Kumar, 1995). Among the most applicable traditional AI techniques are:

- knowledge-based systems – computer programs that represent and apply knowledge and expertise to the more or less (expert systems) specific problem areas
- machine learning – programs that can accumulate and modify their knowledge following new experience
- planning – programs that recognise a final goal and can develop plans for meeting it under a set of given constraints.

Much of the research in AI has concentrated on problem solving. The following section describes an AI powerful problem-solving technique – the knowledge-based system (KBS) approach, and one of its most applicable versions – the knowledge-based expert system (KBES).

Knowledge-based (Expert) Systems

The emergence of knowledge-based systems in the 1980s was the most promising realisation of research in AI. Since then, many research and commercial applications have been developed in areas such as diagnostics, design, planning, education and data interpretation. KBSs (and their more focused and narrower version – expert systems) can be defined as interactive computer programs that solve problems requiring humans' intelligence and expertise. The solution to a problem is usually provided by a system acting as an expert adviser during an interactive session with the user. The system starts asking the user a series of questions in order to define the problem and then searches its knowledge base using an inference engine. Once a good enough solution is found, it is presented to the user with an appropriate explanation of why it was selected and how it was found.

The development of such systems is facilitated very much by use of special software tools called *knowledge-base development environments*, or shells which are often written using LISP or Prolog programming languages. Once developed, the KBES will have, among others, the following distinct components, which distinguish them from other programs:

- user interface, including knowledge acquisition and explanatory modules
- knowledge base
- inference engine.

The user interface not only provides communications facilities to the user, but also allows knowledge acquisition and update during the system operation. The knowledge base contains both facts (declarative knowledge) about a subject expert area (for example, range of productivity rates for different labour gangs) and heuristics, or rules of thumb (procedural knowledge) defining the way an expert applies the reasoning procedures (for example, the 'formwork for foundation' activity to be carried out by gang X in February, the expected rate of productivity is between N1 and N2). The inference engine module processes the knowledge related to a specific problem (domain knowledge) and makes associations and inferences resulting in a recommended solution or course of action. The way the system arrived at these can be traced and explained.

Though KBESs are powerful computer tools offering real solutions in many areas, their application success in planning and design is less impressive than promised. Among the main reasons are a difficulty in finding 'real experts' and acquiring their knowledge, and structuring and representing this knowledge in the computer's memory. These are objective difficulties that results from complicated and ill-structured processes as planning and design. The research community is testing new tools, such as case-based reasoning (CBR) and fuzzy logic techniques as possible solutions.

Object-Oriented Approach

The growing complexity of software systems (for example, AutoCAD release 13 has about 2.5 million lines of code) and the subsequent difficulty in maintaining and extending them (e.g. the Year 2000 'millennium bug' problem) have prompted a search for more efficient and effective tools for software developments. The object-oriented approach (also known as object-oriented technology, or object-orientation) has been introduced as a better methodology for the development of complex software systems. By focusing on representation of data (or knowledge) and operations, the object-oriented approach seeks to (Wix, 1995):

- develop identifiable parts of the software code ('objects') in isolation from the other parts
- make such objects reusable within other software systems.

The key elements of the object-oriented approach, making it different from the conventional approach, are abstraction, encapsulation, inheritance and interoperability.

Abstraction allows simplification of the definition of complex data types as basic components (objects) including methods which can define size, functional behaviour and connectivity. This makes it much easier to model complex products, such as buildings, and related processes, such as planning and design (Bridges, 1995).

Encapsulation means that objects can be used by a program, or integrated in a model without any detail of their structure or internal representation. Similarly, a designer requiring specific output for an air-conditioning system in an office does not need to know *how* it operates.

Inheritance allows the definition of types of objects (classes) as sub-types of other classes. The complete building can be defined as an aggregation of storeys, rooms and structural elements in a hierarchical way. As a result, all columns and beams which belong to a specific storey will inherit its general properties, such as a name or storey number, but they can also have their own properties, such as material, strength and cost.

Interoperability, based on the polymorphism concept, enables objects to pass information to each other and to react in different ways to information received. This feature is considered to be most beneficial from the user point of view. Based on an object's unique identity, the interactions between objects are usually carried by message passing. A message can be sent to a whole class of objects, sub-class of objects, or just a single one. Such a message can simply update data or trigger a complex function. This mechanism allows a complex system to quickly respond to changes, updating the affected objects automatically.

The object-oriented approach has been adopted as a development methodology in research and development projects dealing with integration, product and process modelling, information exchange (e.g. COMBINE, ToSEE, STEP, IFC – Industry Foundation Classes, and other initiatives).

Neural Networks

The neural network approach received its name from the human brain's mesh of neurons. This approach is one of the outcomes from early attempts in the AI branch of computer science research to develop a system which, unlike a 'conventional' sequential and procedural computer, will operate like a human brain, i.e. in parallel on several processes. Indeed, like a brain, a neural network is built from interconnected processors (neurons) which operate in parallel and interact dynamically with each other (see Figure 3.5). Unlike the brain, the mesh has a much simpler structure and many fewer processors than neurons in a brain. However, even such a basic structure enables the network to learn

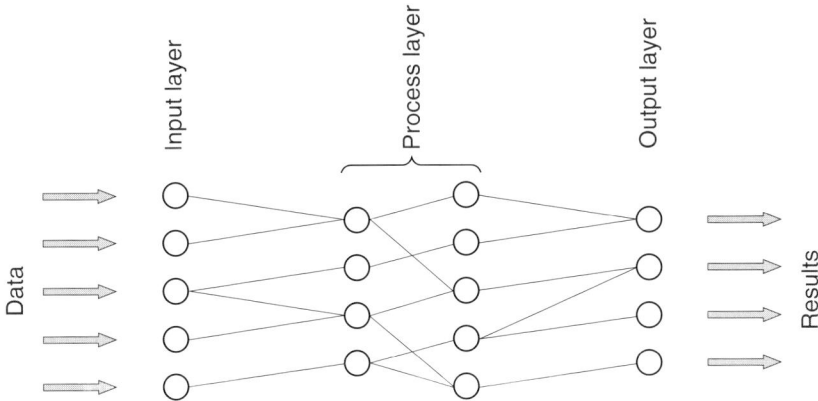

Fig. 3.5 Neural network example

(during special 'training') from data it processes and to adapt itself (through interconnections) to changes in data. It makes neural networks most suitable for the analysis and prediction of generally non-deterministic or highly erratic data with a large number of independent variables, such as image processing, voice recognition, risk analysis, performance prediction, and others. Some examples of neural network applications in civil engineering can be found in Flood and Kartan (1994), Topping (1998) and other sources.

Neural network systems are usually implemented using special software packages. The processing power will be provided by a special board or processor chips ('parallel computing') for 'real-time' or 'heavy-loaded' systems, or can be simulated in software for simpler problems.

References

AZUMA, R. (1997). 'A survey of augmented reality'. *Presence*, Vol. 6, No. 4, pp. 355–80.

BASOZ, N., KING, S., KIREIDJIAN, A. and LAW, K. (1996). 'Visualisation of GIS and network analysis for earthquake damage assessment of transportation systems'. In B. Kumar (Ed.), *Information Processing in Civil and Structural Engineering Design*. Edinburgh: Civil-Comp Press, pp. 151–60.

BRIDGES, A. (1995). 'Object technology in collaborative architectural design'. In Powell, J. (Ed.), *Object Technology and its Application in Engineering*. Daresbury Laboratory, EPSRC, pp. 40–6.

CARGIL, C. (1995). 'The challenge for OGC'. *Geo Information Systems*, May.

BURNS, P., HOPE, D. and ROORDA J. (1999). 'Managing infrastructures for the next generation'. *Journal of Construction Automation*, Vol. 8, No. 6, pp. 689–703.

COYNE, R. D., ROSENMAN, M. A., RADFORD, A. D., BALACHANDRAN, M. and GERO, J. S. (1990). *Knowledge-based Design Systems*. Reading, MA: Addison-Wesley.

DeOLIVEIRA, (1998). 'Case-Based Reasoning in Virtual Reality: A Framework for Computer-Based Training'. Unpublished PhD thesis, University of Salford.

DYM, C. and LEVITT, R. (1991). *Knowledge-Based Systems in Engineering*. New York: McGraw-Hill.

FEINER, S., MACINTYRE, B. and SELLGMANN, D. (1993). 'Knowledge-based Augmented Reality'. *Communications of the ACM*, Vol. 36, No. 7, pp. 53–62.

FLOOD, I. and KARTAM, N. (1994). 'Neural networks: civil engineering: systems and applications'. *Computing in Civil Engineering*, Vol. 8, No. 2, pp. 149–65.

GARRETT, J. and SMAILAGIC, A. (1998). 'Wearable computers for field inspectors: delivering data and knowledge based support in the field'. In Smith, I. (Ed.), *AI in Structural Engineering: IT for Design, Collaboration, Maintenance, and Monitoring.* Berlin: Springer, pp. 146–64.

HERING, J. (1995). 'OGIS and the RDBMS vendor'. *Geo Information Systems*, May.

HOLLOWAY, R. L, (1997). 'Registration error analysis for augmented reality'. *Presence*, Vol. 6, No. 4, pp. 413–32.

KALAWSKY, R. S. (1993). *The Science of Virtual Reality and Virtual Environments.* Reading, MA: Addison-Wesley.

KUMAR, B. (1995). *Knowledge Processing for Structural Engineering.* Southampton: CMP, Southampton.

LARIJANI, C. L. (1994). *The Virtual Reality Primer.* New York: McGraw-Hill.

MACHOVER, C. and TICE, S. (1994). 'Virtual reality'. *IEEE Computer Graphics and Applications*, Vol. 14, No. 1, pp. 15–16.

MILGRAM, P. and KISHINO, F. (1994). 'A taxonomy of mixed reality visual displays'. *IEICE TRANS. INF. & SYST.*, Vol. E77-D, No. 12, December, pp 1321–27.

MILGRAM, P., TAKEMURA, H., UTSUMI, A. and KISHINIO, F. (1994). 'Augmented reality: a class of displays on the reality-virtuality continuum'. *SPIE, Telemanipulator and Telepresence Technologies*, Vol. 2351, pp. 282–91.

MINGUS, P. (1996). 'The future of GIS'. In *Proceedings of the ICE*. pp. 44–50.

OMAN, C. (1996). 'GIS and the Channel Tunnel rail link'. *Proceedings of the ICE*. No. 2, pp. 19–22.

NAVON, R. and RETIK, A. (1997). 'Programming of construction robots using VR techniques'. *Automation in Construction*, No. 5, pp. 400–12.

O'CONNOR, N. (1998). Applications of Augmented Reality in Construction, unpublished research report, University of Strathclyde, Glasgow.

PARKER, D. (1996). 'An introduction to GIS and the impact on civil engineering'. *Proceedings of the Institution of Civil Engineering*, Vol. 114, No. 2, pp. 3–11.

PAULSON, B. C. (1995). *Computer Applications in Construction.* New York: McGraw-Hill.

PIMENTAL, K. and TEIXEIRA, K. (1993). *Virtual Reality Through the New Looking Glass.* New York: Winderest Books.

RAY, C. F. (1996). 'The use of GIS in a major water utility company'. *Proceedings of the ICE*. No. 2, pp. 22–29.

RETIK, A. (1995). 'A VR prototype for visual simulation of the construction process'. In A. Powel (Ed.), *VR and Rapid Prototyping for Engineering.* Salford: EPSRC, pp. 90–3.

RETIK, A. (1997). 'Planning and monitoring of construction projects using virtual reality'. *Journal of Project Management (APMF)*, No. 3, pp 28–32.

RETIK, A. and SHAPIRA, A. (1999). 'VR based planning of construction site activities'. *Automation in Construction*, Vol. 8, pp. 479–83.

RIBARSKY, W., BOLTER, J., DEN BOSCH, A. O. and VAN TEYLINDEN, R. (1994). 'Visualisation and analysis using virtual reality', *IEEE Computer Graphics and Applications*, Vol. 14, No. 1, pp. 10–12.

RICH, E. (1986). *Artificial Intelligence*. New York: McGraw-Hill.

ROORDA, J. (1996). 'Managing buildings for the next generation: asset manager merit using spatial information systems'. In D. Langford and A. Retik (Eds), *The Organisation and Management of Construction. Vol. 3, Managing Construction Information*. London: SPON, pp. 106–122.

SCOTT, P. A. R. (1994). 'Low-cost remote sensing techniques applied to drainage area studies'. *Journal of Institution of Water and Environmental Management*, Vol. 8, No. 5, August, pp. 497–501.

SHE, T. H. and AOUAD, G. (1996). 'Development of an information model for GIS bridge management systems'. In B. Kumar (Ed.), *Information Processing in Civil and Structural Engineering Design*. Edinburgh: Civil-Con Press, pp. 103–111.

SIMMONS, G. (1996). 'Practical applications of GPS for GIS and civil engineering'. *Proceedings of the ICE*, pp. 30–4.

STALKER, R. and SMITH, I. (1998). 'Augmented Reality Applications to Structural Monitoring'. In Smith, I. (Ed.), *AI in Structural Engineering*. Berlin: Springer, pp. 479–83.

THOMAS, D. (1996). 'Implementing GIS across a business' *Proceedings of the ICE*. No. 2, pp. 12–18.

TOPPING, B. (1998). *Neural Networks and Parallel Computers in Construction*. Edinburgh: Inberleith Spottiswoode.

VINCE, J. (1995). *Virtual Reality Systems*, SIGGRAPH Series. Wokingham: ACM Press Books.

VINCENT, S. P. R., METCALFE, R. E. and TONG, D. P. (1996). 'Practical application of satellite remote sensing as a source of information for civil engineering projects and GIS'. *Proceedings of the ICE*, pp. 35–43.

WIX, Y. (1995). 'Object technology: a general view'. In Powell, J. (Ed.), *Object Technology and its Application in Engineering*. Daresbury Laboratory, EPSRC, pp. 1–9.

Chapter 4

Computerised Organisation and Use of Information

'I hear and I forget, I see and I remember, I do and I understand.'

Confucius

Introduction

The construction process is more difficult to computerise than, for example, manufacturing. The main reasons lie in the nature of the process, subsidiary on, and related to, human factors. Furthermore, the process is very long. From the inception of an idea or need to completion of even a modestly sized building or structure can take two to five years. More complex projects will take much longer. The uniqueness of the process, reflecting in bespoke projects, prevents a direct investment of both time and resources into design of a product and planning of its construction. In addition, fragmentation of the process and usual separation of owners, consultants and contractors (either in location or ownership) create fundamental hurdles in integration to all parties involved.

Nevertheless, latest developments in information technology have resulted in both software advances and lower hardware costs. These trends help to overcome some of the difficulties, especially in communication and transfer of information. Access to information sources and design data is much easier with a rapidly developing network infrastructure. All these demand more careful attention to organisation of data and information processing in order to support and facilitate the information access, use, transfer and interpretation.

Chapter 3 introduced fundamentals of data organisation on the basic, or task, level. This chapter presents and examines a systems approach to management of computerised information on higher levels in order to support decision making during planning and design of construction projects.

We will review the information sources, present the information levels and characteristics, and describe a systems approach.

Information Sources

The vast range of information that a construction professional is likely to come into contact with reflects the wide range of disciplines that construction draws upon. So, information rooted in management, technology, engineering, economics, law (contract, environmental, and health and safety), personnel management, operational research and statistics, to name but a few, will be pertinent to the work of construction managers. Not only are the sources diverse but the requirement for information is separated by time. Information may be provided continuously, periodically and occasionally. The differences are delineated by Newcombe et al. (1992): 'Continuous inputs are produced frequently with short time intervals between each provision, e.g. site labour productivity measures; FT share indices.' Periodic information occurs regularly at prescribed intervals. External information such as the Budget and the Housing and Construction Statistics appear periodically, as does internal information such as a company's annual report and accounts. Occasional information occurs irregularly and may be generated externally by, for example, reports on the industry's performance or internally as a reaction to events.

However, the type of information that construction professionals demand will vary according to their position within the organisation, the purpose of the information, and the level of detail required. It could originate from external or internal sources.

External information encompasses reports and similar documents which have been produced using raw data. Examples would be the government's or other organisation's reports on how the national Budget will influence the construction industry, etc. This type of information is of necessity general and individuals and organisations will need to contextualise the information for their particular purposes.

In contrast, internal information is organisation specific. It is information generated and used by an organisation and its managers. At the corporate level, information such as the strategic plan may provide internal information which is at the pinnacle of the information pyramid, but this will frequently be turned into secondary information such as marketing plans. At the project level the master plan or milestone chart may be broken down into monthly, weekly and daily short-term programmes.

The information available to construction professionals is presented in a variety of forms. The main media for delivery of this information are still in the format of paper reports, drawings, journals, texts, etc. Increasingly, information technology and computer-based information sources provide construction information.

Whatever the form or type of information, the level at which it is used may be broken down into three categories:

- the information necessary for an understanding of the construction industry, its structure, organisation and future direction
- the information necessary for the planning and control of firms operating within the industry – be they consultants contributing to design or contractors undertaking construction work
- the information necessary to ensure effective project management including co-ordinated project information.

The main sources for each level are briefly reviewed below.

Information Sources for the Industry

The primary sources of information about the nature of the construction industry is provided by the government. Foremost in the UK is the quarterly *Housing and Construction Statistics*, which trace the economic and employment trends in the industry. These are broken down into the various regions. The range of information available from the data is produced within the Department of the Environment, analysed by the Central Statistical Office and published by Her Majesty's Stationary Office (HMSO).

Government data can also be found in publications such as the *New Earnings Survey*. This is produced by the Department of Trade and Industry and charts the changes in hours worked, basic pay and earnings across a wide range of crafts and professions. This enables construction managers to check for any differences between the wages and salaries a company is paying to that typical of the industry, or other industries, in a region.

Those companies sophisticated enough to engage in formal corporate planning will frequently seek information which is indicative of the business behaviour of potential clients. Gathering data about political, economic, social and technological trends (PEST analysis) can assist a construction organisation to direct its marketing and align the firms' resources to growing, future markets, not current and perhaps dormant ones.

Corporate Information

The big picture provided by the government statistics is infilled by a wide range of information sources which seek to be either futuristic, current or historical in emphasis. Information sources useful to a company can:

- forecast trends in particular construction markets
- describe the current state of trade
- provide an historical account of performance on a particular facet of the construction industry.

Forecasting information

All this information is by its nature general and locates growth or decline in sub-markets. The most notable source for such information is produced by the company JFC and published by the Construction Forecasts and Research organisation. It provides a quarterly commentary and projections for the various market sectors that make up the industry (private non-housing, industrial construction, etc.). Other forecasting agencies include Cambridge Econometrics.

Construction firms will want to know whether a client has intentions to build and if so what its creditworthiness is. Information provided by specialist companies (for example, Trade Indemnity plc) enables firms to examine the track record of other firms when it comes to paying their bills on time. Such information may well influence the tender submitted for particular clients. The issue of the timeliness of payment has been incorporated into the Housing Grants, Construction and Regeneration Act 1996 (the Construction Act).

The current state of trade

Employers' associations, professional institutions and leading journals serving the construction industry all conduct surveys on the current state of trade. By studying the results, the construction manager is better able to prepare and plan the direction of the individual enterprise within the industry.

Historical account of performance

Information that records an aspect of the industry's performance is a useful benchmark for comparing an individual firm against an industry standard. This can be financial data, size by turnover, number of employees, profit/ earnings (PE) ratios, and so on. This is typically produced annually by some construction magazines (for example, *Building Magazine*) and organisations.

Project-based Information

Project information may be layered, according to Day and Langford (1990), into three levels:

  a general information relating to the project, such as:
- building regulations
- standard method of measurement
- product information.

  b specific information on firms involved in the project, such as:
- office procedures
- manufacturing techniques.

    c   project information specific to the project site, such as:
- brief
- scheme design
- production drawings
- bill of quantities or other vehicle for determining price
- construction programme
- as-built drawings
- design and other warranties.

Following the Data Coordination Report (1971), which identified a need for a coherent approach to presenting and handling project information, several of the principal professional and trade organisations (RIBA, RICS, ACE and BEC) established a body entitled the Co-ordinating Committee for Co-ordinated Project Information. After some eight years' work the concept of Co-ordinated Project Information (CPI) was launched. The documentation on CPI included guidance on the layout of information. Central to this concept was the Common Arrangement of Works Section (CAWS), which sought to organise construction project information into 300 work categories based on four source documents: the Standard Method of Measurement; the National Building Specification; the Price Adjustment Formula categories; and the Property Services Agency General Specification. However, the co-ordinating committee for CPI found it difficult to reconcile these documents and so based their 300 work sections on observations of how work is carried out on site. The common arrangement of works information is structured in three levels:

- level 1 is a group of activities, e.g. groundwork, air conditioning/ ventilation
- level 2 is a subgrouping which typically defines the work of a specialist contractor
- level 3 is a description of an operation.

These three levels may be illustrated as follows:

Level 1	Group	Section D Groundwork
Level 2	Subgroup	Section D3 Piling
Level 3	Work Section	D30 Cast in place concrete piling.

Supporting CAWS is the *Production Drawings Code*. This advises designers about the arrangement and presentation of drawn information, including sheet sizes, annotation, titling and numbering as well as the structuring of information. The *Project Specification Code* aims to make project specification comprehensive, precise, practical, specific and clearly written. The specification relies on libraries of clauses based on the library of Standard Descriptions, written by authors from the PSA, the RICS and BEC.

    Under the CPI system the specification is an integral part of the design, not something prepared by the quantity surveyor and semi-detached from the architects' design work. The project specification is organised in accordance with the methodology established for CAWS. CPI is completed by the Standard Method of Measurement 7 (SMM7), an integrating document, set out in accordance with CAWS, to forge coherent links between design information and that needed to build. The SMM7 enables bills of quantities to be prepared in SMM order and so be co-ordinated with other project information using a coding system that creates an appropriate environment for computerisation.

Project information sources for designers

Most designers will have access to catalogues of trade literature compiled and updated by the premier construction project information source: Barbour. The *Barbour Index* holds all the trade literature relevant to buildings. The RIBA product information is a similar project index.

Project information sources for contractors

Much of the project information held by contractors arises from information generated by other participants in the project. But contractors will draw on standard databases for cost and price data and production information to build up estimates for work and construction programmes.

Computerised Information Sources

Computer networks and telecommunication systems not only increase accessibility of and provide user friendly retrieval facilities to data banks, but they also become more cost effective. As a result more computerised information sources are available today. Some use computers to facilitate collecting, inputting, maintenance (update) and retrieval of the existing sources of information (e.g. technical indexes); others create new tools in order to increase effectiveness of information management (e.g. knowledge-based systems, multimedia approach, visualisation, telecommunication and networking).

Today's software packages allow users to produce in-house databases and information systems tailoring the information use and management to the organisation's format and convenience. Usually access to such information is denied to the 'rest of the world'. The concepts, development and use of the information systems are presented later in this chapter. This section will describe the information sources produced by public and commercial agencies for external users. These sources (usually called databases or data banks) can be delivered to the user's computer *off-line* (on disk, tape, CD ROM, etc.) or *on-line* (via computer or telephone networks). The major advantage of having an on-line service is the almost instant access to the latest information. On-line information retrieval is carried out using a computing terminal to search for data (textual or numerical) stored on computers around the world. Search terms can be combined in various ways on-line to give a more specific search than is possible using printed volumes. The cost of use depends on the database searched, the amount of time spent on line and the number of data items printed.

Saur-Bowker, an international publisher, periodically publishes a comprehensive review on *Information Sources* for different disciplines. The latest issue of *Information Sources in Engineering* (1996) includes in its 'Construction management' chapter some examples of computerised databases (Langford and Retik, 1996).

Off-line information sources and services

Off-line information is usually a part of the task-orientated commercial software (mainly on diskettes) or large on-line databases (mainly on CD ROM). The most updated information (local or national) about these sources can be found in software catalogues, directories and professional journals, and at computer exhibitions, etc. Nevertheless, there are several directories of publicly available databases available world-wide. *Directory of Portable Databases* and *Computer-Readable Databases* (both by Cuarda/Gale) are among the most comprehensive.

The following are some examples of such databases:

- *RIBA.ti* – Construction Information Service is available on CD ROM or as a network version. The database includes technical information and the Building Supplement (equivalent to more than 150 000 pages of text) such as British Standards, Codes of Practice, Legislations, Agreement Certificates, etc.
- *CSSP/Thomas Telford Software* (owned by the Institution of Civil Engineering, UK) develops and markets commercial database management systems for the construction industry. Standard libraries for quantities measurement and price databases are integrated in the software and can be altered by the user.
- *Means* (R.S. Means Company, Inc.) is a large and comprehensive source for US construction data. The company publishes over 20 annual cost guides and about 50 reference books. All cost data are available in Lotus 1-2-3 format or through different commercial estimating software packages.
- *PERINORM* contains European standards and technical regulations.
- *Wessex* (Wessex Software UK Ltd.) produces commercial software for estimating, pricing of bills of quantities, etc., integrating construction cost databases from price books published by Wessex Group. The

software enables the databases to be altered either globally or by estimate to allow users to adjust data to suit local conditions and individual prices.

On-line information sources

There are over 10 000 (ten thousand!) on-line databases available worldwide. Several commercial companies provide current information on databases accessible through on-line services.

Gale Research publishes as a part of the *Electronic Information Series* several directories covering various aspects of the information industry. Among them is the *Directory of On-line Databases*. It provides detailed entries with database type, subject, producer, on-line services, product description, language, geographic coverage, time span covered, and frequency of updating. Separate sections list contact information for database producers and on-line services, as well as master, subject and geographic indices. Another important publisher is the Association for Information Management (Aslib), which publishes the *Aslib On-line Series* (see Cox, 1991). The series includes more than ten rather subject-orientated directories.

A unique example of a comprehensive on-line database is *BCIS on-line* (Building Cost Information Service, owned by the Royal Institute of Chartered Surveyors). It has been designed to allow sections of data from the BCIS data banks for storage on the user's own computer. The data bank now has over 10 000 cost analyses from UK construction projects available to subscribers. Analyses covers a period of 20 years from 1973 but the majority (73%) are from the last ten years. Indices, construction economic indicators, background information as well as the Approximate Estimate Package software program are also available to the subscribers. These are updated weekly.

Among other examples are *ICONDA* (the CIB International Construction Databases). Its bibliographic section only contains about 300 000 references. *Scan-A-Bid* and *TED* (*Tenders Electronics Daily*) provide daily information on the international bids and tenders.

Electronic networks and major on-line hosts

There are many on-line databases which are accessible through the educational and commercial networks of the Internet. For example, JANET (Joint Academic NETwork, which connects campus networks in the UK) provides access to bulletin boards (BUBL for services resources available on networks), specialist databases (bibliographic BIDS, software NISSPAC, HENSA) computerised catalogues of academic institutions (UK, US/Europe) and Email.

Information services companies (called network hosts) provide on-line directories of the databases available from them. *Data-Star*, the largest European on-line host system, provides users with access to over 250 public databases. *DIALOG Information Services* is a huge database host offering access to over 400 public databases worldwide. Other big hosts where construction information can be found are *BRS Search Services, ESA-IRS, FIZ Technik, Knowledge Index, Orbit Search Service, PFD On-line* and others.

Information Levels and Characteristics

The purpose of information is to assist planning, control and decision making on operational, tactical and strategic levels.

Operational information is information that will help project consultants, managers or site supervisors to organise, perform and control the tasks for which they are responsible. This is information which is generally more short term in nature – e.g. daily, weekly or perhaps monthly in timescale. Examples of operational information are:

- availability and cost of materials
- the weekly figures for the section of the site for which the supervisor is responsible
- staffing levels; that is, the numbers of staff expected (full-time and part-time), details of absenteeism, lateness in arriving at work, etc.
- the date of expected materials deliveries on site.

Tactical information is information that will help middle management to plan and control the resources of the business for which they are responsible. Accounting information of a tactical nature, for example, includes the cash flow and monthly performance reports. Ideally, tactical information should include both a target for achievement and actual results for comparison against the target.

A project manager might be interested in the following tactical information:

- details of actual monthly performance
- income/expenditure report in a form of cash flow
- cost of running different divisions/departments (plant, materials, etc.).

Strategic information is information that will assist management to make strategic planning decisions, most of which will affect the longer term future of the organisation. Strategic planning decisions are those concerned with ensuring that the organisation achieves its objectives which might be expressed, for example, in terms of profit and assets growth.

Examples of strategic information are:

- a long-term profit forecast
- capital expenditure requirements over the next few years, potential sources of finance to fund this expenditure requirements over the next few years, potential sources of finance to fund this expenditure

- information about companies that might be suitable take-over targets or potential predators.

The characteristics of information management are as follows:

(i)   It should have a clear purpose. Without purpose, there is only 'data' not information. The manager should clearly realise what the purpose of the information is.

(ii)   It should be accurate and precise enough for the purpose in hand. There is often no need to produce values to the nearest £1, for example, when dealing in amounts of thousands or millions of pounds. When providing information for planning or decision making, where uncertainty about the future is an integral factor in the situation, there is no point in providing figures which pretend to have a degree of precision that simply would not be possible.

(iii)   Where values are uncertain, some information giving assessments of probability distributions or sensitivity analysis might be particularly useful.

(iv)   Information must be comprehensive but relevant. Managers cannot take reliable decisions without all the necessary relevant information.

(v)   Information must be sent to the proper person − i.e. the manager who is responsible for the decision in hand. It is valueless if it is sent to a person who lacks the authority to act on it. The level of detail in the information should be suited to the position of the manager in the organisation hierarchy. In general, junior managers need more detail, and senior managers less.

(vi)   Information should be timely. It ought to be available for the time when the planning or control decisions are to be taken (or ought to be taken). At the same time, it should not be provided too frequently. Excess information, in terms of both quantity and frequency, is wasted and time-wasting.

(vii)   Information must be clear to the user. It can be accurate, timely and relevant, but if the user does not understand it (e.g. accounting reports may contain too much jargon for non-accounting managers) it will not be useable.

(viii) Where possible, information should be concise. The principle of reporting by exception is particularly useful for control reports, but even planning reports (e.g. reports to a board of directors with a recommendation) can often benefit from brevity. Lengthy reports take time to read and digest, and might obscure the essential points with much less essential detail.

It will be worthwhile producing more information if the marginal benefits expected from using it exceed the marginal costs of providing it. The costs of obtaining extra information or more detailed or more frequent information could be assessed through careful study. The marginal benefits are less easy to establish, although some approaches can be used to estimate the expected value of both perfect and imperfect information.

The rules in communicating information to management are as follows:

(i)   The information should always be as accurate and as precise as the recipient would like it to be. The degree of precision required will vary according to circumstances and the project stage.

(ii) The information should be comprehensive and complete. The recipient will be dissatisfied with having only a fraction of the information that the recipient does not understand, perhaps because too many jargon words are used.

(iii) The information should be clear. It is useless to send information that the recipient does not understand, perhaps because too many jargon words are used.

(iv) The information should be sent so that it is received in good time to enable the recipient to use it. Information that is received too late is useless.

(v)   The information should all be relevant to the needs of the recipient who will not want to wade through items of irrelevant data searching for the information that is actually required.

Information Systems

The term 'system' is probably used more in IT environments than any other word (i.e. computer system, expert system, CAD system, and so on). Yet it is not always used in the same broad way as the 'information system' concept means. Therefore, some background on both concepts and fundamentals of information systems is required in order to comprehend, analyse and manage construction processes and is presented below in the following sections.

Information Systems Concepts

The system concept is a generic one and therefore can be found in almost all areas of human activities. Indeed, we talk about social systems, biological systems, weapon systems, manufacturing systems and even our Sun system.

In all these, a system is a group of interrelated or interacting elements forming a unified whole. Moreover, the following definition provides a better framework for describing information systems (O'Brien, 1993):

> A system is a group of interrelated components working together towards a common goal by accepting inputs and producing outputs in an organised transformation process.

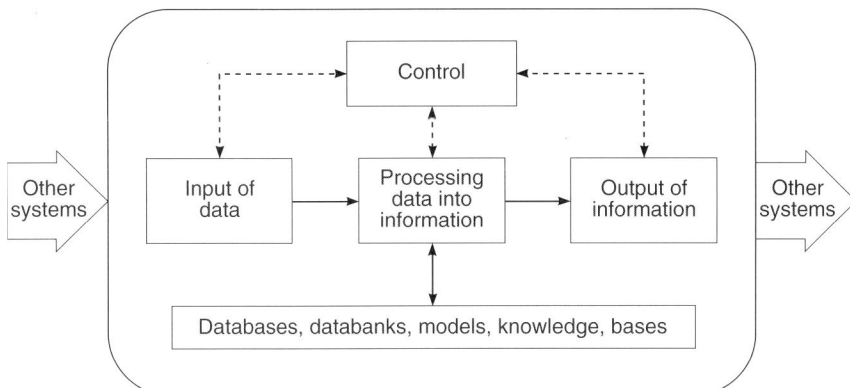

Fig. 4.1   Information sytems

The information system model in Figure 4.1 presents major activities of an information system. The four basic components involved in performing these activities (compare to 4M resources in construction) are people (*m*en and women), hardware (*m*achine), software (*m*oney) and data (*m*aterial).

Information System Components

Four of the above presented components or resources are fundamental parts of any (manual or computer-based) information system. All are equally important for the system performance and will be described briefly below.

People

People, or human resources are those who either develop and operate or use information systems. Development and operations of information systems are carried out by specially educated or trained specialists such as system analysts, programmers, computer operators, technical and managerial personnel. Information system users (often called end users) are people who use the system for their professional activities (e.g. engineer) or decision-making support (e.g. managers).

Data

The word 'data' is the plural of 'datum', though it is quite usual in database literature to use it in both singular and plural forms. Data are a series of facts or raw material of information systems. Data can be alphanumeric (numbers and alphabetical characters, e.g. cost data, calculation tables, etc.), textual (written text, e.g. specifications, standards, etc.), geometrical (drawing), graphical (pictures, maps) and even audio and video.

Computer Integrated Planning and Design

All data in information systems are organised and held in databases, which may have different structures and be related to different models for their representation and calculation. Many novel information systems will have knowledge bases to facilitate data interpretation and even to give advice to users.

Hardware

Hardware in a computer-based information system will consist of:

- computer(s) processing the data
- peripherals, supporting input, output and storage activities
- telecommunications media interconnecting users and system components within and outwith system environment.

Software

Software includes:

- programs (sets of operating instructions to direct and control hardware)
- procedures (sets of information processing instructions to be used by people).

Processing of Data into Information

Information systems will have input, processing, output, storage and control activities in order to produce information required. These activities are also called data or information processing. Though 'information' is defined as processed data, the terms 'data' and 'information' sometimes may be interchangeable. For example, data on quantities and types of ready-mixed concrete supplied by a construction company concrete plant will be processed into information on the total sales in that day. Such information from every plant will provide input data to the company's information system. Once the data processes, the information on the company's sales that day can be obtained and compared with previous days. Interpretation of results and subsequent decision making will require appropriate *knowledge* and experience. Whatever the level of information processing, it usually takes the following route:

- Data are either collected manually (i.e. special forms), or captured by using special devices (e.g. scanners, data cards) and prepared for entering into a computer system by recording, usually onto magnetic disks on tapes. The user can access and edit the recorded document through a user interface.
- The data then can be processed by organising, analysing and manipulating into required information. The information can be

delivered to the user as a message, chart, report, printed document or graphics display.

Not only input data but information and its product can and should be stored and made available for later reuse or retrieval. What data and information to keep and for how long are not simple decisions. Subcontractors' accounts and transactions will probably be kept for a few years after completion of a project, while for time schedules only one previous version is important for actual-versus-planned comparison.

One of the computer paradoxes is that people too often take for granted computer results as correct. Therefore, the appropriate 'control mechanism' has to be built in a system in order to provide feedback about all its activities. The feedback should easily be monitored and understood allowing any mistakes (as a result of incorrect input) to be captured immediately.

Globalisation of Information

There is a growing trend for companies operating internationally to move toward a transnational business strategy. As a result, we all face many cases where products, say, were designed in Europe, manufactured in China, assembled in Mexico and sold in the USA. Similar cases are also seen in construction, for example Hong Kong Airport (Schmitz, 1996), where companies all over the world were involved. This trend needed IT support and triggered development of hardware, software and telecommunications technology to produce global information tools. These tools were designed to facilitate for the companies the integration of the global business activities of their subsidiaries and headquarters; additionally, they had to deal with national restrictions on transborder data flows, movement of personnel, availability of hardware and software, difficulties in development standard procedures and common data definitions.

However, although the impact of information technology on business activities and developments is usually positive, the impact on ethical and social issues should be carefully examined. Issues such as employment, working conditions, computer monitoring, privacy, health, computer crime and others should be properly considered and addressed.

Those issues that seem to be more relevant for the construction companies will be looked at in the following subsections.

Virtual Organisations

Communications infrastructure developments, such as global area networks and mobile telecommunications networks, combined with high reliability and lower cost of communications services created two important trends –

globalisation of companies and flexibility of individuals. Examples of the latter are 'working from home' employees, 'hot-desking' and the 'anywhere office' approach implemented by British Airways in their Waterside office. Also, the emergence of a new breed of sophisticated and informed customers has increased demand on industry and organisations to deliver better value, higher quality and innovative products and services. Existing models and concepts of organisations and management are not always capable of responding to these challenges and opportunities (Dulaimi et al., 1996). The new organisation model has emerged: the 'virtual organisation' or 'virtual company'.

The virtual organisation is an alliance of organisations and/or teams of individuals coming together and sharing skills, expertise and knowledge to respond to specific demand, opportunity or challenge. Such organisation is usually comprised of groups of professionals separated by geographical distance, time difference and psychological and even social barriers. These groups use communications technology to join together. They can be temporarily engaged in several virtual organisations working on different projects or towards a common venture. Since virtual organisation is not bound (meanwhile!) by the traditional roles governing employers and employees, it can harness the power of market forces to develop, manufacture, market, distribute and support their offerings in a way that traditional companies cannot duplicate (Cheshbrough and Teece, 1996). Virtual organisations can quickly reshape their core and supporting members in response to client demands and environmental and even political conditions (Guss, 1996).

Cheshbrough and Teece (1996) and others found that virtual organisation approach has some similarities with procurement – the practice of construction projects, which brings together human skills experience and expertise as well as physical and material resources from external and often distributed organisations. Though virtual organisations can benefit from experience and problem-solving approaches for co-ordination and management of construction projects, creating virtual environment networks would not be a simple task for the construction industry itself.

Guss (1996) describes some examples of videoconferencing and Internet technologies used for communication between partners of construction projects. The results of a survey for one of the projects (see http://www.gcn.net/arena) show that parties involved are very satisfied with teamwork, co-ordination, communication of problems and response time. However, there are a number of fundamental differences from traditional project procurement and management practices that create challenges for implementing virtual organisation within the industry. These challenges surprisingly, are not necessarily technical (the communications technology already provides quite adequate facilities) but social and organisational. These are results of traditional project management focus on co-ordinating fragmented technical tasks rather than integration of technical and communication processes (Guss, 1996). Moreover, the virtual organisation concept is based on much higher concurrency of tasks than even those experienced in design–build, BOT/BOOT (build–operate–(own)–transfer) and similar types of procurement techniques. Possible solutions to these and other problems are outlined in Chapter 5.

Social Aspects

Information technology combined with globalisation aspects affects not only organisational but also social and ethical issues such as the impact of IT on employment, working conditions, privacy, individuality, health, computer crime and others. All these are results of information technology tools giving us ability to acquire, manipulate, store and communicate information in seconds to practically any person in any place.

Those people responsible for business decision making should also be aware of the ethical aspects of information technology use, to prevent misuse of this powerful technology. For example:

- Should the computer technology be used to monitor your employees' work activities?
- Should you electronically access the employees' personnel records?
- Should you sell or transfer information about your customers, extracted from company transactions?
- Should you allow employees use of their work computers and copies of software for their private use?

It is obvious that the general ethical philosophies of behaviour and business ethics (Langford et al., 1995) should be adhered to. However, information technology often has additional aspects and impacts that deserve a close look.

IT, employment and working conditions

As far as employment is concerned, IT can have adverse and/or beneficial effects. Full computerisation of a design office or automation of a construction materials factory will probably reduce demand for new jobs or even eliminate some existing ones. However, the productivity and quality should be improved at less cost as well as creating job satisfaction for employees. Moreover, computerisation usually improves working conditions forcing upgrade of the work environment and investment in infrastructure. The calculation power of computers shifts the content of work itself from repetitive and tedious activities toward more challenging and creative tasks. As a result, IT contributes to improving skills and education, creating a culture of training and learning. Also, new jobs, such as for programmers, IT managers, etc., are created in many places. The ethical role of a manager in such an environment is to maximise beneficial effects, keeping the negative effects to the minimum.

IT and human factor

Many early computer systems were not 'user-friendly' and flexible. This often created a negative impact on individuality of people, treating them as numbers on records and not being able to find a person responsible for a mistake. 'It's a computer fault' could justify any mistake and allow those responsible for it to have 'a machine to blame'. On the other hand, many software interfaces were not 'built' properly for intensive use, as well as forced people to work the 'computer's way'.

Ergonomical engineering today concentrates on both hardware and software design, creating more 'people-oriented' computers, where many human factors even such as humour, are seriously considered (O'Brien, 1993).

The increased use of computer technology created many situations where people sit at personal computers or workstations almost all of the working day. Such intensive use triggers a variety of health problems, such as damaged arm and neck muscles, eye strain, radiation exposure, job stress and others (Betts, 1990; Dejoie et al., 1991; Dunlop, 1991), Solutions for some of these problems often lie in work environments that are comfortable, pleasant and safe, and proper job design, which provides work breaks every few hours and variety of tasks. Ergonomics, or human factor engineering, is a branch of science addressing these issues.

IT and privacy issues

Information on individuals, very often confidential, is collected and contained in many computer databases of government and private organisations. There are cases where the information has been stolen or misused, resulting in fraud, invasion of privacy and similar offences. Moreover, even unauthorised use of such information or errors in records could damage the privacy of individuals.

In many countries privacy acts and other related laws strictly regulate the collection and use of personal data by government agencies and private companies. These acts specify not only duties of organisations to safely keep, maintain, update and protect data, but also individuals' rights to inspect and correct their personal records.

Computer crime

Computer crime is use of computer and information technology for criminal or irresponsible activities. It is a rapidly growing threat, with a worrying tendency of many criminal organisations taking advantage of IT to support their 'usual' activities (Ramo, 1996). Moreover, computer technology has created its own offenders and criminals, such as Internet hackers, not all of whom are harmless teenagers who break into the computer networks for a challenge (Anon, 1998). Thus, the Model Computer Crime Act of American Data Processing Management Association defines computer crime as one of the following:

- the unauthorised use, access, modification and distraction of hardware, software, or data resources
- the unauthorised release of information
- denying an end user access to his or her own hardware, software, or data resources
- using or conspiring to use computer resources to illegally obtain information or tangible property.

Money and services theft are the most 'popular' types of computer crime. Unauthorised software copying, called software piracy, is a widespread form of software theft. Software is intellectual property which is protected by copyright law and user licensing agreements. Payment for a software package gives a licence to use the software. Paying for a site licence means permission to legally copy and use software within a particular location ('site') or an organisation by its users. Public domain software is not copyrighted and can freely be copied and distributed.

Computer viruses and worms are also examples of computer crime. Though many of them are harmless or just annoying jokes, others have a very disruptive and destructive intention of destroying the data, software, contents of memory, hard disks and other 'malicious' actions. A virus is a small 'invisible' program that can work only when inserted into another program, unlike a worm program that can run and act on its own.

Both viruses and worms, usually recognised and referred to as viruses, 'infect' computer systems through copies of software, files or network links.

References

ANON (1998). 'Computer security and the Internet: special report'. *Scientific American*, October, pp. 69–89.

BETTS, M. (1990). 'Repetitive stress claims soar'. *Computer World*, 11 November.

CHESHBROUGH, H. W. and TEECE, D. J. (1996). 'When is virtual virtuous? Organising for innovation'. *Harvard Business Review*, January/February, pp. 65–73.

COX, J. (1991). Building, construction, architecture databases. London: Aslib.

DAY, A. and LANGFORD, V. (1990). 'The implementation of CPI in the UK building industry'. Report to the Building Centre Trust, *University of Bath*.

DEJOIE, R., FOWLER, G. and PARADICE, D. (Eds) (1991). *Ethical Issues in Information Systems*. Boston: Boyd & Fraser.

DULAIMI, M. F., BAXENDALE, A. T. and LANGFORD, D. A. (1996). 'The organisation project as a process of innovation'. In Langford, D. A. and Retik, A. (Eds), *The Organisation and Management of Construction*, Vol. 2. London: SPON, pp. 572–7.

DUNLOP, C. and KLING, R. (Eds) (1991). *Computerisation and Controversey: Value Conflicts and Social Choices*. San Diego: Academic Press.

GUSS, C. (1996). 'Virtual teams, project management process and the construction industries'. In *Proceedings of CIB W78 Workshop on Construction on the Information Highway*, June, 1996, pp. 253–64.

HOUSING GRANTS, CONSTRUCTION AND REGENERATION ACT (THE CONSTRUCTON ACT) (1996). London: HMSO.

LANGFORD, D. A., FELLOWS, R., HANCOCK, M. and GALE, A. (1995). *Managing People in Construction*. Harlow: Longman.

LANGFORD, D. A. and RETIK, A. (1996) 'Construction Management'. In Mildren, K. and Hicks, P. (Eds), *Information Sources in Engineering*. London: Bowker-Saur.

NEWCOMBE, R., LANGFORD, D. and FELLOWS, R. (1990). *Construction Management*. London: Batsford.

O'BRIEN, J. A. (1993). *Management Information Systems*. Boston, MA: Irwin.

RAMO, J. C. (1996). 'Crime online'. *Time Digital*, September, pp. 16–20.

SCHMITZ, J. (1996). 'Programme management of the Hong Kong Airport projects'. *IPMA'96 World Congress on Project Management*, Paris, June 1996, pp. 629–38.

Part 2

Applications of IT to Construction Projects

Chapter 5

Computer-aided Project Planning and Scheduling

'A person who knows how to solve problems is always less efficient than a person who knows how to avoid them.'

Genaro Cons

Introduction

The planning and scheduling processes form the basis of project management today. The planning procedures and scheduling techniques are well established and comprehensively described in many sources (Callahan et al., 1992; Harris and McCaffer, 2001; Illingworth, 2000; Pilcher, 1992; and others). Most construction firms use formal scheduling procedures whenever the complexity of work tasks is high and the co-ordination of a multiplicity of different operatives and trades are required (Hendrickson and Au, 1989). On construction sites, in addition to assigning dates to project activities and tasks, scheduling is intended to facilitate the matching of resources of equipment, materials and labour with project work tasks over time. Good scheduling can reduce or eliminate bottlenecks and facilitate the procurement of critical activities, thereby ensuring timely completion of the project. All these facilities, if carried out manually, can be time consuming and labour intensive.

Sometimes the enormity of the project would render efficient project planning and scheduling unfeasible if it were not for the computerisation of the planning and scheduling duties. With the continued development of computer software and improved graphical presentation media, many of the practical problems associated with formal scheduling mechanisms have been overcome. Although it takes considerably more than a computer and some project-management software to manage projects effectively, the advent of project-management programs has revolutionised the practice of project management and has assisted project managers in expediting their duties more effectively than they have done in the past. Some of the functions involved in project management, especially those concerned with project control, e.g. cash-flow monitoring, risk and delay management, were difficult to execute efficiently or quickly enough before computers were used.

First, this chapter reviews existing computer-aided planning and scheduling tools designed to support project management. Then, new developments are presented and future research and development trends are outlined.

Current Tools for Project Management

The rapid growth in the power of microcomputers has made it possible for construction managers to effectively and efficiently analyse the massive amounts of data necessary to monitor and control the progress of the many interrelated tasks that go together to make up a construction project. This facility has enabled professional and technical staff to spend a greater proportion of their time on specific project-related tasks rather than on the mechanistic administrative tasks that can now be carried out more efficiently by computer. Microcomputers have become commonplace tools in assisting project managers, planners and schedulers with the complex and time-consuming calculations involved in determining schedule dates and other information related to the scheduling process.

The growing use of microcomputers has resulted in an unprecedented increase in the development and supply of software tools designed to fulfil specialised requirements. Accountancy, job costing and CAD are a small selection of specialised functions for which software developers are providing specialised products. The functions of project management, planning and scheduling have also attracted the attention of software firms. As a result, there are more than one hundred different project-planning software packages on the market today. What is the difference between them? What benefits can these tools bring? The following section attempts to answers these questions.

Characteristics of Software Packages

In order that the software and its relevance to project management can be discussed it is necessary to frame a checklist which will assist in its evaluation. The evaluation criteria used in this instance fall into three basic categories:

General characteristics

The commercial name of the system, its list price, the name and address of the software house, and the sort of work that it is designed for.

Technical features

Including features such as graphical presentation, modelling, resource assignment, resource scheduling, multi-project capability, tracking, cost allocation, report formats, data transfer capability and so on.

Specialist features

Features that can be used in the field of delay and claim management and which can be used in addition to the basic features required by most planning and scheduling software packages.

Several surveys and studies of commercially available packages were carried out within the last several years in the Department of Civil Engineering, University of Strathclyde (Barr, 1997; Conlin and Retik, 1997; McLellan and Mansfield, 1993). The results of these studies form the basis for the following analysis.

When microcomputer project-management software first reached the marketplace, it seemed that the products could be grouped into one of two distinct categories. At the lower end the software was simple to use but provided very little functionality. At the higher end, on the other hand, it provided functionality but was too sophisticated for any but those who were already using mainframe project-management software; and, when compared with the mainframe, meant a reduction in speed and capacity and the inability to share a database. The majority of users wanted the functionality of the high-end programs with the ease of use of the lower end products. As technology has moved forward, the gap between these two ends of the program spectrum has diminished. This has resulted in a wide range of software providing a wide range of capabilities and addressing the needs of most project managers.

Although a continuum still exists, it is useful to subdivide the continuum into four main groups to facilitate ease of understanding. This classification into groups has been made to permit the definition of individual requirements, allow the development of unique criteria for selecting software, and identify some of the software which can be used to meet this criteria.

Base Level

There must always be a low end of any range (Levine, 1989). When examining project-management software it is useful to classify this level as the entry level. Although this level of software can be seen as very basic it does not mean it is useless. Every program on the market has a function for someone. It is only by comparison with the high-level programs that the entry-level programs appear deficient.

In studies recently completed (Barr, 1997; Conlin and Retik, 1997), the examination of project-planning software packages has been limited to mass-market packages, advanced packages and sophisticated packages. The base level, for the most part, has been excluded from the study as it was felt that packages in this classification were not applicable to construction-related situations.

It is sufficient to say that the majority of base level scheduling packages are little more than computerised appointment books, or they are purely graphical tools with little or no database facilities. The former classification has tended to transform itself into the personal information manager classification whereas the latter has transformed itself into the graphical or charting package which allows attractive, easy-to-read charts and graphs to be drawn by keystrokes or by mouse.

The base level package does not really lend itself to a construction environment although if the situation requires a mouse drawn bar chart without the data normally associated with higher level software packages, then a base-level package may be all that is required. It has previously been the case that such a situation, although in the construction environment, does not require graphical presentation. A system which is commonly used by construction clients is the pure database such as dBase. Many clients are only concerned with a few key dates in the construction process and are generally not interested in the comprehensive and detailed presentation of a contractor's or engineer's project plan or schedule. In such cases, the database software can provide all the data manipulation that is required.

Table 5.1 contains comparison of mainstream software packages. Each product has been allocated a category within which it can be described and evaluated. These categories are denoted as follows:

- MM = Mass Market
- Ad = Advanced
- So = Sophisticated

The packages were placed in these categories according to their apparent functionality and the type of work they are intended to carry out.

Mass-Market Level

In terms of this study the mass-market level of software packages represents the greatest part of the software under consideration. The distinction between the mass-market level of programs and the base level is not great. Mass-market programs are usually characterised by their attempt to provide 'a little something for everyone' (Levine, 1989).

The mass-market software packages differentiate themselves from the base-level packages by offering the user the facility for handling more activities, by permitting more user-defined options, and by allowing, albeit in a limited mode, the user to perform certain control functions. Packages in the mass-market category are ones which provide the user with a set of tools and facilities which will allow them to carry out the most fundamental project-management functions: for example, to enter a certain level of cost data, to allow resources to be allocated to tasks, to provide a limited range of pre-determined reports, to allow the user to specify what type of view is required.

It must be stated that price is not an indication of the category to which the software package is assigned, nor is it an indication of how powerful the package is. Notwithstanding this, there certainly appears to be a remarkable consistency in the pricing of the mass-market products. The prices of these products range from £100 to £2,500 with the majority being priced around the £700/£800 mark.

Included in the mass-market level are some very familiar products names: for example, Powerproject Horizon, CA-Superproject, Hornet XK, Instaplan, Artemis Schedule Publisher, Pertmaster, On Target, Project Scheduler. These products, on the whole, provide a certain level of technical features that are commonly used in the construction industry.

Advanced Packages

Advanced packages are identified by their ability to offer the user the facilities of packages at the mass-market level but which also allow the user to carry out more project-control functions. These functions include the ability to allocate variable resources to tasks, the ability to allocate different types of cost data to tasks, much more powerful user-defined reporting formats and the ability to carry out limited project-tracking operations.

Packages at this level also offer more sophisticated features. There should be sufficient capability to build an accurate model of the project and to track results against that model. For this level of functionality the user will pay a price. Packages in this market sector range from £1,500 to £3,800 with the majority of them costing more than £3,000.

The advanced software packages really attempt to emulate the functionality and capabilities of the sophisticated mainframe-level packages whilst still trying to provide the user with a package which is easy to use and understand.

Sophisticated Packages

Sophisticated packages are defined by the fact that they allow the user to create a bespoke system for a particular project. These packages allow the user to carry out full project-control activities including full project tracking and updating.

The programs in this, the top group, offer most of the functions available in project-management software. As a group they tend to permit highly

Table 5.1 Features of project management software

Product	Vendor	Tasks per project	Resources per project	Multiple calendars	Materials and costs	No. of views	Resource loading reports	Shows resource loading conflict	Resource levelling	Marks critical path	Macros	Dependencies	Lan	Additional languages	Category
Powerproject Horizon	ASTA Development Corp.	200000	10000	Y	N	5	Y	Y	Y	Y	N	N	Y	3	Mm
CA-Superproject	CA Limited	Unlimited	Unlimited	Y	Y	7	Y	Y	Y	Y	Y	Y	Y	9	Mm
Hornet 5000I	Claremont Controls Ltd	5000	128	Y	Y	0	Y	Y	Y	Y	Y	Y	Y	Fr&S, p	Ad
Hornet XK	Claremont Controls Ltd	5000	128	Y	N	5	Y	Y	Y	Y	N	Y	Y	Y	Mm
Instaplan	Deepak Sareen Associates	Unlimited	Unlimited	Y	Y	7	Y	Y	Y	Y	Y	Y	Y	Fr	Mm
Primavera (P3) 5.1	Forgetrack Ltd	Unlimited	Unlimited	Y	Y	0	Y	Y	Y	Y	N	Y	Y	5	Ad
Artemis 7000	Lucas Management Systems	999999	999999	Y	Y	0	Y	Y	Y	Y	Y	Y	Y	Y	So
Artemis Schedule Publisher	Lucas Management Systems	12000	Unlimited	Y	Y	0	Y	Y	Y	Y	N	Y	Y	Y	Mm
Microplanner V6	Micro Planning Ltd	1364	26	Y	Y	0	Y	Y	Y	Y	N	Y	Y	Fr&G	Ad
Microplanner Profession	Micro Planning Ltd	Unlimited	26	Y	Y	0	Y	Y	Y	Y	N	Y	Y	None	Ad
Pertmaster Advanced 2.4	Project Management Shop	Unlimited	Unlimited	Y	Y	0	Y	Y	Y	Y	N	Y	Y	None	Mm

Product	Vendor												
Project/2 Series X	PSDI Inc	Unlimited	Unlimited	Y	Y	0	Y	Y	Y	Y	Y	Y	Ad
On Target v. 1	Symantec (UK) Ltd	1500	Unlimited	Y	Y	0	Y	Y	Y	Y	N	Y	4	Mm
Time Line v. 5	Symantec (UK) Ltd	1500	300	N	Y	0	Y	Y	Y	Y	Y	Y	5	Mm
Project Schedular 5	Tekware Ltd.	2000	500	Y	Y	0	Y	Y	Y	Y	N	Y	Fr&G	Mm
Open Plan	Welcome Software	Unlimited	1430	Y	Y	0	Y	Y	Y	Y	Y	Y	Fr G&M	So

Legend

Ad = Advanced level, Hd = Hard disk, Lan = Local area network, H = Hercules, Mm = Mass market level, So = Sophisticated level, Fr = French, Sp = Spanish; G = German, M = Mandarin

Notes

Unlimited: Unlimited in this instance means theoretically unlimited.

Data: All data contained in this table has been abstracted from the manufacturers' specifications obtained in the vendors survey.

detailed modelling of the project workscope and of resource and cost assignments. They all feature resource allocation and levelling, progress tracking, and user control over report content. The two programs included in this group are Artemis 7000 (from Lucas Management Systems) and OpenPlan (from Welcome Software Technology). These are the kinds of programs that are attractive to project control people who are used to working with mainframe-level packages or who need the functionality that such software offers. These programs are widely used in the construction industry. The price range starts at £1,295 whilst the most expensive costs around £5,000.

Specific Features

The broad range of available software is apparent by a cursory look at Table 5.1 and the categories described above. In order that the correct software package is chosen for a particular work the following features also have to be taken into account.

Data input

Data input is, perhaps, one of the most important features for a mass-market product. A product which is clearly targeted at this market has to provide a facility which is so easy to use that a beginner can, essentially, pick up the product and obtain output immediately. On the other hand, it must not present a too simplistic image, otherwise it will not appeal to experienced project managers familiar with project-management software.

One of the most popular features of packages in this market is the compose-on-screen facility. This facility allows the user to create a bar chart or network on the screen as if he/she were drawing it on paper. Some analysts are of the opinion that this feature is initially attractive to new project-management software users but that it could prove slow and tiresome to experienced users. The experienced user may find this particular feature a nuisance because it requires more key strokes and screen movement than some alternative solutions. These alternative solutions range from inputting to a form on the screen (one activity at a time), inputting to an input table on the screen (multiple activities), or inputting to an external file (or a database) that is then read by the project-management program (batch inputting).

Project tracking

This feature is probably one of the most important when examining the role of computerised project-management software in time-based construction claims. All the packages in the mass-market category allow the user to prepare a project plan with varying degrees of cost and resource information. For complete project tracking, though, the packages must offer the user the capability of freezing a project plan, including the schedule, resource plan and budget. Once this has been done the user must be able to enter actual start dates, actual completion dates, actual resource use and actual costs without changing the original data. This function allows the user to compare the actual results with the original intent. This facility is at the heart of the as-built method, time impact analysis method and as-planned method of delay and claim analysis (Conlin and Retik, 1997).

Networking mode

Another function which is important when selecting scheduling software for a particular application is the networking protocol. The term 'networking protocol' means the method by which the user defines relationships between tasks. It is important to select a software package which allows the user to specify complete precedence relationships: for example, start-to-start, finish-to-finish, finish-to-start, start-to-finish, as well as specifying lead/lag durations.

Resource planning

Resource allocation and costing are important features in any project management and scheduling situation (East and Kirby, 1990). This importance is increased considerably when the user requires the software in a situation where control or modelling of resources and costs is necessary to prepare cash flow forecasts, or to mitigate the effects of a delay or claim.

When choosing a software package considerable care should be taken to ensure that the resource-based functions meet the criteria stipulated by the situation which the package is being considered for. It is not advisable to choose a package on the strength of a vendor's sales literature or even on the strength of short reviews. This is because, even after taking the above details into consideration, there can be minute differences which are important to the user. This is especially true of software packages which are likely to be employed in delay and claim situations.

Cost planning

In basic terms there are two primary ways in which project-management software packages approach cost planning. The program can be written to deal with costs for each activity, which are determined by the definitions of the resources and their unit costs, or else the program can allow only costs that are fixed for each activity as a specific input rather than being calculated by the resource quantities and rates. In the resource–times–rate approach, the program usually also allows fixed-cost inputs for materials and equipment . Each of these approaches has variations in the way in which they operate.

Macros

This feature is frequently overlooked by purchasers when considering a potential software package. The ability to write, retain and use macros effectively can greatly increase a user's speed, efficiency and accuracy when using a package. This is especially the case when the user carries out a certain action or activity on a regular basis.

Macros can be written to do virtually anything that the user could do with a number of keystrokes within the software package. Macros can play a very important part in the field of delay analysis and monitoring, especially when the user wishes to obtain maximum accuracy in his or her actions.

Reports

This is another very important factor to consider when purchasing a project-management software package. The whole point of having and using a piece of powerful and expensive software such as those shown in Table 9.1 is to provide and present reports to personnel or management who can then study the information contained in them and implement some form of action. If the package does not produce reports there is not much point in having the software at all as it will become just another pretty graphical package. The ability to produce reports is what distinguishes the base-level 'personal organisers' from the heavy-duty project-management packages.

Data integration

The ability to link up with other software packages is, perhaps, one of the most important features in today's data-transfer environment. The capability of exchanging data with spreadsheet packages, database-management systems, statistical packages and ASCII files is so important that it requires special mention. Also included in this discussion is the capability of project-management software packages to import and export data to/from word-processing applications for the purposes of including in reports, graphics and/or data lists.

The data-link feature is, so often, overlooked when a user is considering purchasing a software package. This should not be the case. To extract the best from any project-management software package it is necessary to have the facility to talk to other packages, especially ones which are recognised as industry standards, such as Lotus 123, MS Access and dBase.

Advanced IT Tools

The majority of commercial project-management applications are restricted to carrying out their tasks by a process of formal numerical analysis of relatively well-defined problems. The techniques and applications discussed in this chapter form the base for new types of computer-based aids for project management. Amongst those discussed in more detail are knowledge-based expert systems, visualisation and VR, simulation and telecommunications.

Knowledge-Based Expert Systems

A knowledge-based expert system (KBES), or its more focused and specialised variation – an expert system, is a computer program which provides advice in a selected, specialist field. It uses specific knowledge of a particular area, called a domain, and heuristic problem-solving methods to perform functions normally reserved for a computer expert (see also Chapter 7).

The success of any expert system depends on the system developer's ability to formalise and represent the knowledge and problem-solving procedures employed by a particular expert. Hendrickson and Au (1989) recognised that this expertise may consist of the ability to recognise a particular situation or pattern in the environment out of many thousands of possible solutions, and that this pattern recognition is difficult to emulate in a computer program. There are a few applications of KBESs in the field of construction management. KBES developments are described in research literature. Though they are very seldom developed into stand-alone commercial applications, the contribution to managerial knowledge and other software development is valuable. Dym and Levitt (1991) suggest that the management of construction projects will still require many uniquely human skills, although KBESs for planning, scheduling, monitoring and control will act as valuable decision support tools for construction professionals.

From what has been discussed previously, it can be seen that knowledge-based expert systems can play an important part in the construction process. At different levels of complexity the KBES will play varying roles. Tasks which are routine and repetitive could become fully automated whereas tasks which require substantial input from humans will use KBES technology in addition to the human input.

Planning by Simulation

Simulation is a powerful tool in analysing and providing a possible solution to a problem. It allows the testing of various operational options before problems even occur and without conducting live experiments. It is not surprising, therefore, that there has been a rapid growth over the last few years in computer-simulation software. Such software allows setting up and conducting simulation not only by expert analysts but also by practitioners.

Computer-based simulation can be defined as 'the process of modelling and representing the behaviour of a system on a computer' (Cotton and Oliver, 1994). The simulated system can be a model of a real or artificial system, process or product. Therefore, when the simulation is taking place, the system's behaviour and responses can be observed and analysed in different situations. Moreover, if a simulation model is dynamic (i.e. incorporates the element of time), it could be built to represent a long period of time by a short-duration period.

The simulation can be of *continuous* (i.e. flight simulators) or *discrete*, (i.e. queuing system) type. It can use *mathematical* (i.e. linear programming, regression analysis) or *statistical* (i.e. random event, Monte Carlo) techniques to model a dynamic situation. A simulation model can be *deterministic* (if its behaviour is entirely predictable), *stochastic* (its behaviour cannot be entirely predicted as in the case of random events, probabilistic variables and so on) or *steady-state* (if the probability of being in any particular situation is always the same). General examples of simulation applications and their benefits are described in Pidd (1992), Robinson (1994) and others.

Sensitivity of construction projects to many factors (both internal and external) on the one hand and a high level of uncertainty on the other, make it very difficult to link strategic design-and-management decisions while evaluating alternative strategies during planning. Stochastic simulation models and other probabilistic approaches have more successfully used to produce a production plan or cost–time forecast than deterministic approaches (see Green and Gray, 1997; Halpin, 1990; Tommelein and Zouein, 1993). Such models, however, although generally fulfilling their intended purpose, are known to have limitations when applied at the preliminary stages of a project. The most significant of the limitations are:

- An operational plan for the construction work must be prepared. At the critical, formative stages of the project the knowledge necessary to formulate such a plan is not normally accessible and the conventional planning techniques employed are detailed and time consuming.
- The key aspects of the management control during the construction process are not represented in the systems which rely on the stochastic evaluation of a predetermined plan. The reality of construction projects is different. A construction project involves extensive tactical planning by the site management during the production process together with management action. In the real world the stochastic element is not allowed to run unchecked, but triggers responses from management.

Hybrids of simulation and knowledge-based systems offer a potential solution to these problems. A knowledge-based approach can be used to support the formulation of the production plan upon which the simulation will be based. This will speed the setting up of the simulation to a level where 'what if' experimentation can take place, and it would enable users with limited expertise in contract planning to undertake the process. The potential is there to develop tools well suited to the early decision-making stages.

A knowledge-based approach can also be used to provide the system with the ability to emulate the monitoring and control actions of site management through run-time interaction with the simulation. Under this control action, simulation runs should more accurately represent feasible real-time situations. The dispersion of simulated outcomes should therefore be a better estimate than just a distribution of deterministic outcomes.

A Knowledge-based Simulation Approach to Cost–Time Forecasting

This section describes a recently completed study which investigated the feasibility of an intelligent simulation approach to construction cost and time forecasting (Retik and Marston, 1995). During the study a pilot system, capable of acting as a concept demonstrator, was developed.

The purpose of the system is to facilitate the evaluation of alternative strategies under uncertain conditions. As indicated in Figure 5.1, an approach is envisaged that will enable the user to test alternative management strategies and design requirements outputting probabilistic estimates of project cost and duration. The research was limited to works involving internal modernisation of low-rise housing: these projects exhibit the characteristics of multiple-unit housing rehabilitation in general, yet provide a manageable scope for pilot system development.

The schema in Figure 5.1 serves to illustrate the conceptual structure of the approach. A knowledge-based module, the *auto-planner*, formulates a production plan for the project according to the description of the work required (*design*) and strategy to be employed in its execution (*management strategy*). This *project plan* determines sequence, planned duration and planned resource levels, and acts as the model of the production process during simulation. The *simulator* calculates 'actual' durations, drawing randomly from probability distributions which represent uncertainty of the real world present in actual site performance. The knowledge based *auto control facility* emulates the control actions affected by site management during the production process, monitoring the simulation and triggering actions such as activity acceleration or re-scheduling. The *estimating module* calculates cost on the basis of the resources required in the simulated production

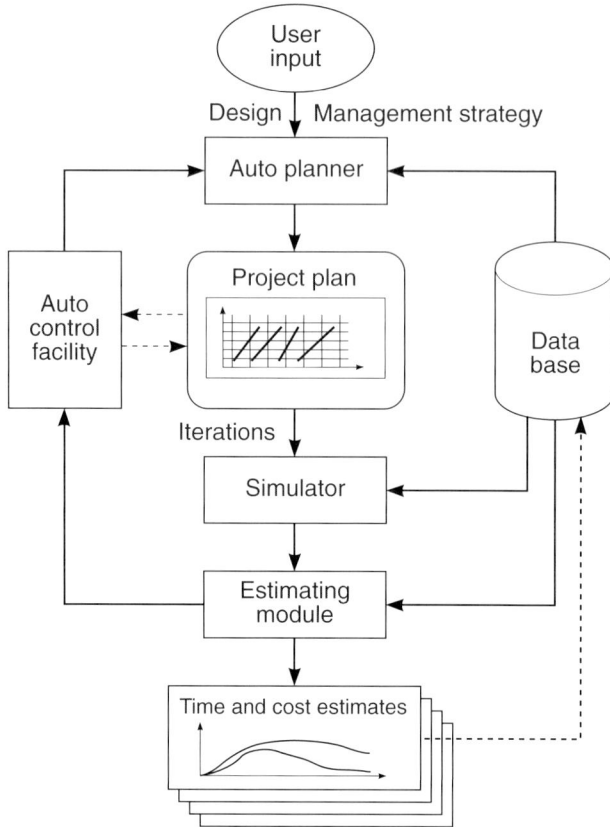

Fig. 5.1   Conceptual structure of the system

run. The full production process is simulated repeatedly (*iteration*), the results providing a *probabilistic estimate* of the project outcome in terms of *cost and time*.

Knowledge-base Structure

A hierarchical structure has been developed which connects high-level user input about the proposed works through to low-level technical detail. By this means strategic decisions about the requirements and context of the proposed works are linked to resource consumption. The structure comprises four object class levels:

A   The **Project** level describes the general class of the works (for the prototype only one type is under consideration, internal modernisation). It also carries information which is common to the whole project.

B   The **Elements** level describes the project in terms of groups of activities. The main requirements of the project in terms of work to be done are specified in instances of this object class.

C   The **Activities** level includes all works that are likely to occur within a particular element – in effect a kind of library of activities. By using information inherited from the higher levels and predefined rules, the required activities are selected. Knowledge about possible sequences of the works and their influence on project duration (sequence, criticality and so on) is contained within the slots as default values and may be modified by a set of rules and functions according to the context in which the work is undertaken. This level is active not only in setting up the project plan, but also during simulation. The data necessary for duration calculation of activities in the planning stage (gang sizes, productivity, quantity of work, etc.) and 'actual' values of productivity in the simulation stage are passed up from the lower 'Items' level.

D   The **Items** level allows the connection of activities to their component resources to be used for cost and duration calculation. The slots contain values of the parameters of the distributions of productivity rates and factors representing the effects of context variables such as work stage, season, conditions and such like. All items may be updated automatically by connection to an external database, or interactively by changing the slot values within the **items**.

Rules and Procedures

There are four main categories of rules and procedures in the system:

1   Plan content selection. This set of rules selects the activities required by the project and determines their sequencing from a predefined set of technological precedents.

2   Plan information generation. This consist of a set of rules for establishing the pace (rate) of working and required gang sizes. Messages and procedures are used to create and update slot values, as well as for the automatic creation of instances for the 'Activities' level. The main goal of this set of rules is to establish a realistic pace for the works by taking into account desirable gang sizes and organisational constraints.

3   Duration and cost calculation. Rules and procedures apply the line of balance technique. In the planning stage, the same working pace is applied to all instances of activity. In the

simulation stage, simulated 'actual' durations are substituted for the planned duration of each activity in each workplace.

4   Decision making in response to monitoring of the construction process. The rules here cover situations where the planned schedule or resourcing requires updating and/or change as a result of simulated 'actual' performance.

System Operation

Figure 5.2 gives an outline representation of the system as a process. There are two main stages of system operation, separated in the figure by a dotted horizontal line; first the generation of a planned schedule and second the simulation of site production. The main procedures are as follows:

Input Project Description

Common project information and description (such as project name, place, number of houses) as well as decanting (tenants in/out during the project) and managerial (handover rate, compliance with planned duration/cost) strategies are entered into the system interactively by the user. Project content, which in the preliminary stages is defined as a work programme only (the kind of work to be done), is entered in the next stage.

Break-down into Activities

The breaking down of work descriptions into activities is a process of a heuristic rule-based selection of an appropriate set of activities within each 'Element' group for a given context of work. These rules are based on establishing technological solutions, identified through knowledge elicitation.

Establishing the Sequence and Pace of Work

Activity sequence and criticality are preset to default values for the full possible range of modernisation work. These values may be modified by sets of rules and functions to take account of activities that are not present in a particular project (knowledge acquisition showed that in practice very few changes in work sequence and criticality result) and the context in which the work is undertaken. This produces an operable rather than optimal sequence, but the user may inspect and overwrite it at will.

Input	Process	Output
Project description and work content	Break down into activities	
	Establishing the pace of the work	
	Duration and cost calculation	Planned schedule
Random productivity	Schedule updating according to 'actual' productivity	Updated schedule
User	Interference and delay identification	Visualisation
	Decision making	Report
	Cost–time Calculations	Probabilistic distribution of cost and time

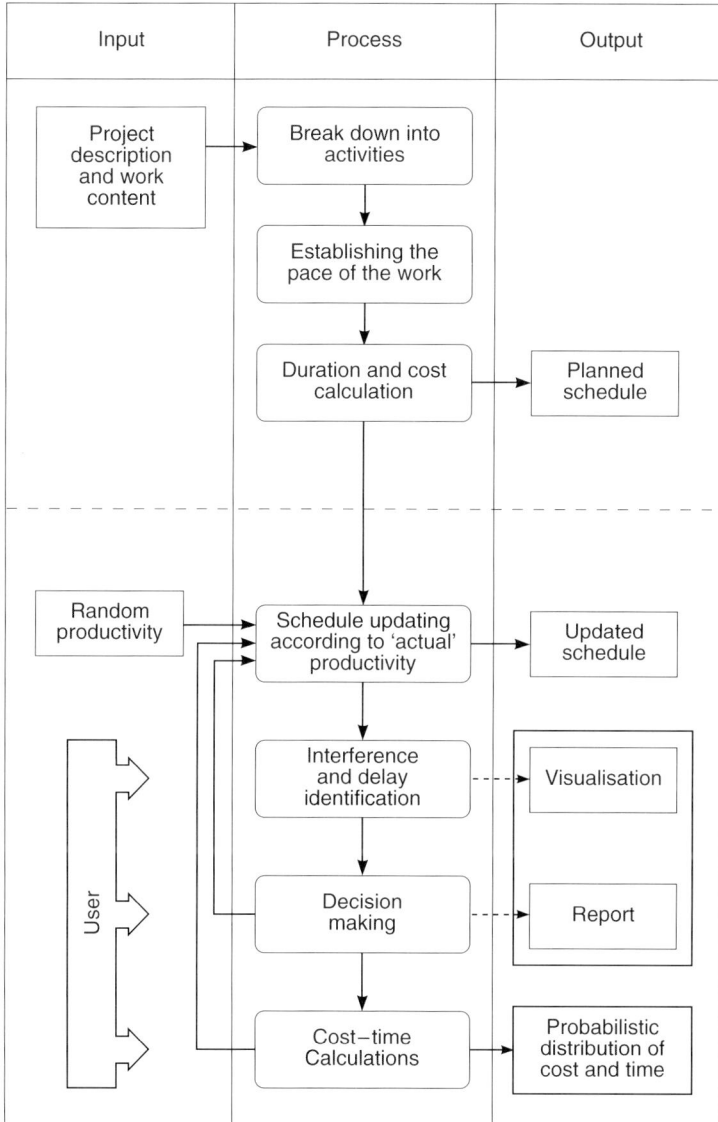

Fig. 5.2   System operation

Cost Calculation

The project cost is calculated in this case as a sum of the preliminary costs, the costs of subcontracted work, the sum of material costs and labour costs for all the activities carried out by the contractor, and the total costs of temporary accommodation.

The Simulation Process

The principal simulation stages are:

1   The 'actual' duration of each instance of each activity is determined by random sampling of productivity rates, following the planned operational sequence.

2   At the completion of each activity in each workplace status checks compare 'actual' with planned to detect interferences and/ or delays. When a significant delay (a default value of which can be modified by the user) or interference occurs, a decision is called for (see site management decision emulation below).

3   At the end of each cycle (simulation of a complete project production run), total cost and duration are recorded.

4   After an appropriate number of cycles (possibly 100), results are reported for cost and duration as statistics and distribution plots for the sample of runs executed.

The parameters of the distributions from which productivity rates are drawn are modified according to the specific project context.

Site Management Decision Emulation

The possible decisions are dictated by the chosen management strategy with regard to the relative importance of cost and time. This strategy is fundamental to the operation of this part of the simulation and must be set at the outset of system operation. From the knowledge acquired to date it appears that the strategy most widely adopted in practice is to set compliance with planned duration as the primary objective rather than cost (see Figure 5.3). This is because of the particular sensitivity of this kind of project to delays: subcontractor continuity, meeting material supply dates and decanting/occupancy arrangements are all of critical importance.

With the time compliance strategy currently in place, response to delay takes the form of increased gang size and attendant cost, triggered when preset tolerances are breached and accruing from the time of the decision. Response to interference involves similar acceleration provisions, but is applied to both the interfering and affected activities.

Presentation of Results

The primary outputs of the system are probabilistic estimates of total project cost and duration, presented both graphically and numerically. Various additional outputs can be produced for the purposes of

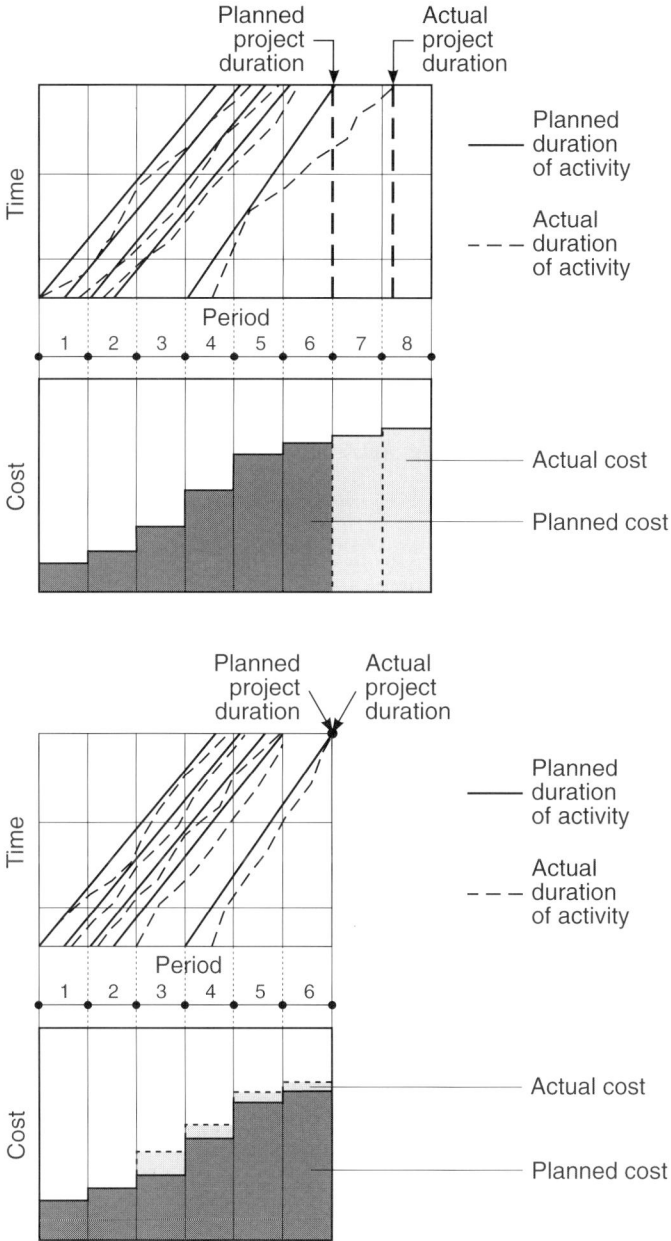

Fig. 5.3   Managerial strategy: (a) compliance with budget cost; (b) compliance with planned duration

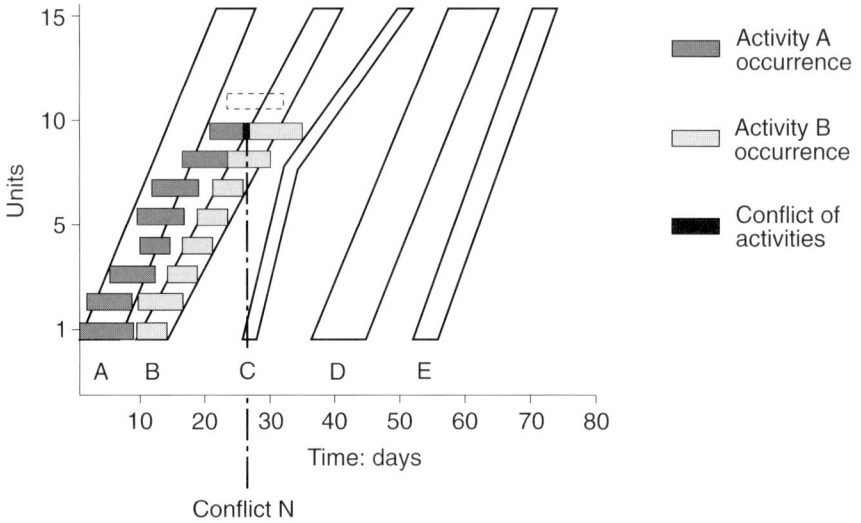

Fig. 5.4   Example of 'post-mortem' analysis

Conflict N:

   Decision: Increase the gang size of the activity A
                  by one labourer for a week

   Implication: add to the 'extra cost' slot the following value:
                  40 hours. Special labour at cost per hour–£12 per hour

explanation and validation including a graphical representation of the operation plan, reports on the decision responses to interference and delay, and a facility to produce a trace of any simulation run so that post-mortem analysis can be undertaken, as represented hypothetically in Figure 5.4 (for an implemented example, see Retik and Marston, 1995).

Visualisation and Simulation using VR

The main contribution of KBES and simulation techniques in project planning lies, as presented above, in decision-making support during either plan formulation (i.e. in order to reduce project duration, avoid delays, optimise resources), or project execution (i.e. progress and cash-flow monitoring, conflict resolution for claim analysis). Similarly, 3D computer graphics are already applied as an aid for visual simulation of a construction project's progress (CST, 1993; Fischer, 1997; Retik et al., 1990). The ability to show a real picture of the work, indicating deviation while simulating alternatives to the planned schedule of project execution, is valuable for the managerial staff in order to identify changes or 'delay-prone' events in the construction process.

Other applications of visualisation techniques can already be found in the courts where the ability to explain your point showing the consequences of delay, for example, to a non-expert in a very short allocated time is often vital. These cases are rarely publicised for obvious reasons; only a number of computer bureaux which advertise these services can provide a possible insight on the 'success' of this kind of application.

Recently, experience of advanced visualisation tools, such as virtual-reality software, combined with increasing power of personal computers has created a fruitful environment for more advanced construction applications (Bridgewater, 1993; Oliviera et al., 1997; Powell, 1995; Retik and Hay, 1994; Ribarsky et al., 1994; Whyte, 1998).

A generic system has been developed by the Virtual Construction Simulation Research Group at the University of Strathclyde. It visualises in 3D a schedule of work simulating the construction process and on-site activities (Adjei-Kumi and Retick, 1997; Retik, 1996). The main goal of the system is to support the planning stage of the construction project, providing a tool to help in 'trying' different methods of project execution (see Figure 5.5). Ability to check buildability of the design as well as to co-ordinate both on-site and subcontractors' activities will significantly reduce inefficient resource deployment and the number of delays (Retik and Shapira, 1999).

Moreover, once a project is in the monitoring stage, the system can compare 'actual' versus 'planned' schedules, highlighting places where delay (if any) occurred (see a demo at http://www.strath.ac.uk/Departments/ Civeng/conman/vcsrg/). Such a presentation of 'problematic areas' can not only facilitate conflict solving but can also help in returning a project to schedule. The project managers who assessed the benefits of the system by testing their own projects, particularly emphasised the schedule information communication potential; the main reservation involved the skills and time required to set up a simulation.

Telecommunications Technologies

Videoconferencing has already been used by some construction companies to interconnect their remote offices. For example, Balfour Beatty (a large UK-based construction company) has linked seven of its UK offices by a videoconferencing system which allows two-way conversations to take place via a 28-inch television screen. The system also allows documents to be transmitted simultaneously to a second screen at the remote end for viewing or printing. Though there are still some meetings to which staff have to (and some want to) travel in person, the overall travel expenditure has been reduced by 20–25% a year (Collier, 1997). Another technology which is becoming more popular among project managers is the closed-circuit television system (CCTV). However, often the lack of appropriate

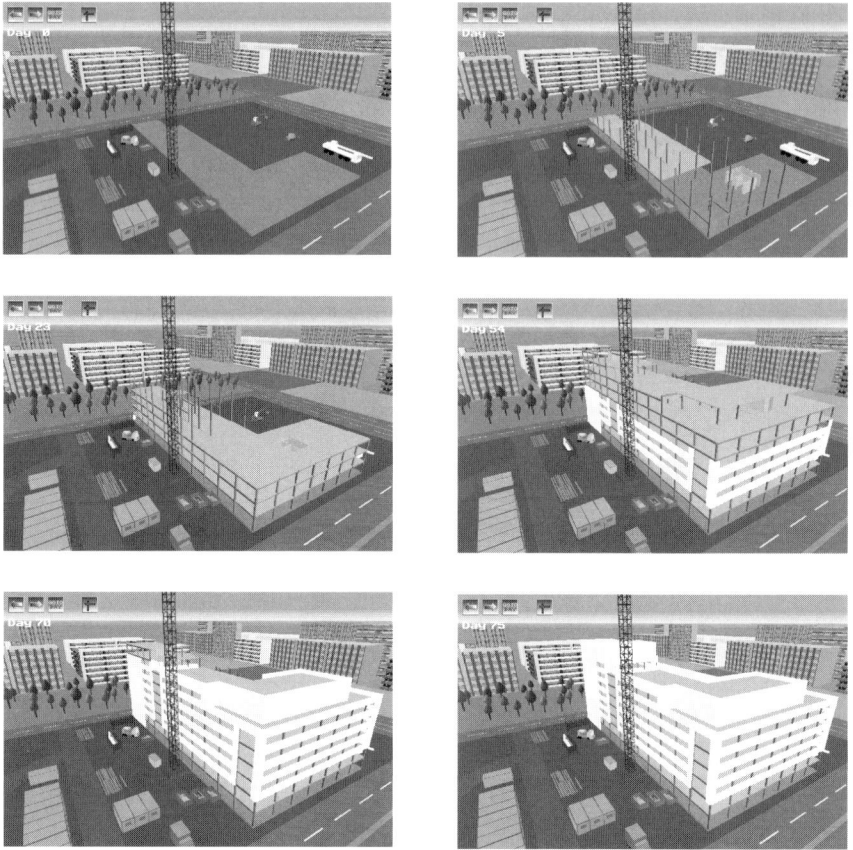

Fig. 5.5a   Visual planning of work in progress: project level

communication infrastructure (such as ISDN lines or cables) prevents the use of these technologies, especially in the case of remote construction sites.

Trying to solve this problem, several research groups have been developing mobile communication systems. One such development is the multimedia hard hat (MMHH) (Thorpe et al., 1995) from a team at BICC plc and Loughbourough University. A site worker wearing the MMHH is able to view plans and information held electronically on a small hat-mounted display. Design office personnel can monitor developments on site through images sent by the video camera mounted on the hard hat and boosted by special radio-relay transmitter.

Another group, based at Strathclyde University (Retik et al., 1997), has avoided the use of an additional transmitter by using specially developed compression software which allows transfer of live video using ordinary mobile phones. This novel research project which integrates the virtual-reality

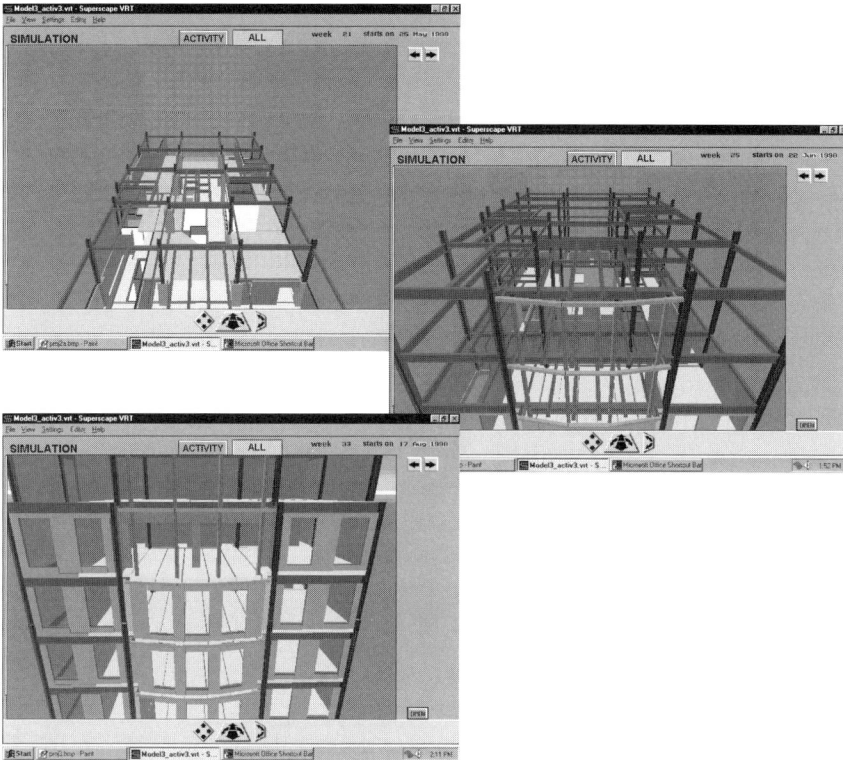

Fig. 5b   Visual planning of work in progress: activity level

(VR), telepresence (TP) and mobile video telecommunications technologies has been carried out by a multidisciplinary team from three departments at the University. In the project context, VR gives the impression of participating in a computer-generated project environment, which is also simulated in time, while TP gives the impression of being present at a remote construction site. A mobile, real-time, 3D-hybrid VR/TP system is currently being built. Once fully completed and tested, it will permit the user to automatically integrate telepresence images with computer-generated virtual environments superimposed over the remote real-world view. This integrated system incorporates emerging mobile telecommunications technologies to give rapid and easy access to the real and virtual construction sites from arbitrary locations. It also allows remote surveillance of the construction site, and integration of real-world images of the site with virtual-reality representations, derived from planning models, for progress monitoring.

The key elements of the approach are:

(a)   *Integration of real and synthetic images* (also called augmented reality): The intent is to allow real images to be integrated into the 3D world.

Such an ability will be crucial for the design and planning of construction projects, which are always an integral part of a project environment.

(b) *Access to remote sites*: Both the project progress monitoring and site control can be performed remotely and examined interactively providing more efficient use of the manager's time. Comparison using intelligent superimposition of 'actual versus planned' situations will create possibilities which are not available today, for example remote site inspection and progress verification by the client.

The conceptual scheme of the system is presented in Figure 5.6.

The camera platforms and the associated control system are being developed from the 'Mark I' Strathclyde Telepresence System by the University's Transparent Telepresence Research Group (Mair et al, 1998). An illustration is shown in Figure 5.7. The new system has already been operated over long range using two cellular phones on a number of occasions. It has unlimited range and is able to operate from battery power thus making it truly mobile.

Visualisation of progress of construction requires representation of changes in the geometry of a building or structure. Such changes prevent efficient use of existing 3D architectural and structural models for planning purposes (Retik, 1996). Moreover, geometrical representation of construction activities is not always 'compatible with' or even presented by design models.

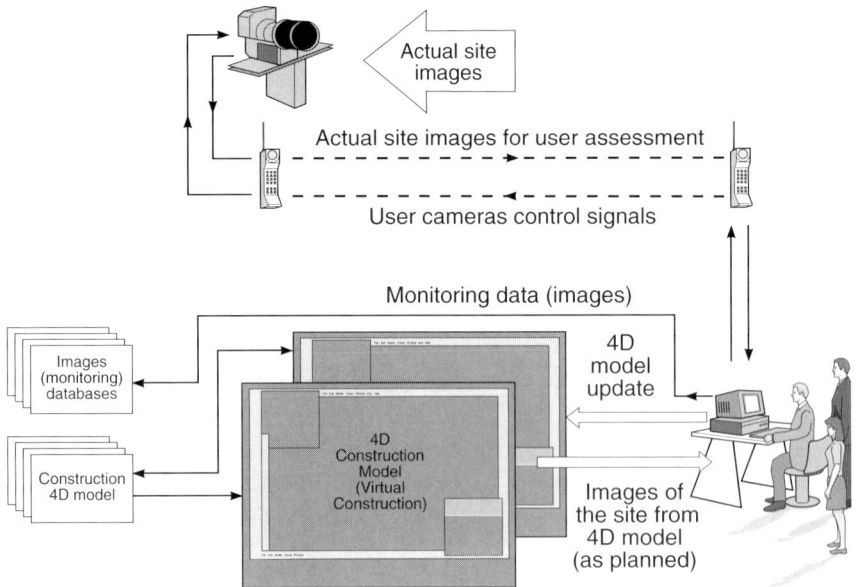

Fig. 5.6   *Interactive visual monitoring of a remote site — a system concept*

Fig. 5.7a   Telepresence system: static version

97

Fixed location
pan, tilt and zoom
camera platform

L.E.O.S
(alternative)

Cellular phone
base station

Mobile vehicle with stereo pan,
tilt and roll camera platform

Fig. 5.7.b Telepresence system: mobile version

Therefore, integration of real images of a site with simulation of, and interaction with, the construction process will require the ability to model dynamic changes to the site geometry.

Integration of real and synthetic worlds is commonly known as hybrid VR or augmented reality (AR). In our case, one of the possible ways of superimposing actual (real images) over planned (virtual images) construction work progress is demonstrated in Figure 5.8. At this stage the images are calibrated and superimposed manually. However, the complexity of construction sites creates difficulties even for simple superimposition and overlapping cases during time-based simulation of the construction process.

Fig. 5.7.b   continued

Fig. 5.8 'Actual vs planned' project monitoring: (a) project level; (b) activity level

Issues such as organising and management of video information, hierarchical representation of the project, image-based retrieval, 3D matching and viewpoint calibration may well require use of advanced AI techniques in order to produce efficient solutions.

Future Trends

The rapid growth in the availability and power of microcomputers, coupled with their continually decreasing cost, has made it possible for construction managers to effectively and efficiently analyse the massive amounts of data necessary to monitor and control the progress of a construction project. Computer-based scheduling takes a central role in this process.

One of the most important features required in planning is, therefore, the ability to produce and manipulate a schedule. This ability is at the heart of resource allocation, project expenditure and cash-flow diagrams, and delay and claim prevention using project-management software. Another requirement of project-management software is the ability to produce comprehensive reports and graphical illustrations. This feature is important because the personnel who are using the software package must be able to communicate their data to other personnel involved in the project. The package must allow the user to analyse and manipulate the data to produce specific situations allowing maximum flexibility in displaying them.

It appears that at this stage in the development of KBESs there is not much difference between a sophisticated package being used in a particular situation and an expert system developed for the same situation. The KBES technology would provide the greatest assistance in the scheduling process when it is linked to a suitable scheduling software package. The specific knowledge required in the field of planning and management would still be supplied by the construction professional although it would be augmented by KBES technology.

This situation will, undoubtedly, change as simulation techniques become more sophisticated and better developed. It should be possible in the future for a construction manager to plan a project using a project-management scheduling software package; then to model the outcome of a project by employing KBES technology; and to simulate the project's execution visually. By doing this the manager can create 'what-if' scenarios, thereby highlighting actions or non-actions which could be the source of potential problems (such as delays and related claims). This does not preclude advanced packages and mass-market packages from being used. It is very much dependent on the users' requirements. If the user does not require the functionality and associated complexity of the sophisticated package then an advanced package or mass-market package would be sufficient.

References

ADJEI-KUMI, T. and RETIK, A. (1997). 'A library-based 4D visualisation of construction processes', *Proceedings*, IEEE Information Visualisation, pp. 315–21.
BARR, G. (1997). 'Current planning and scheduling software tools and approaches'.

Unpublished dissertation, Department of Civil Engineering, University of Strathclyde, Glasgow.

BRIDGEWATER, C. (1993). 'Computer-aided building design and construction'. In Warwick, K., Gray, J. and Roberts, D. (Eds), *Virtual Reality in Engineering*. London: IEE, pp. 65–90.

CALLAHAN, M. T., QUACKENBUSH, D. G. and RAWINGS, J. E. (1992). *Construction Project Scheduling*. London: McGraw-Hill.

COLLIER, A. (1997). 'The end of business travel as we know it'. Science and Technology Supplement, *The Herald*, Glasgow, 25 February, p. 27.

CONLIN, J. and RETIK, A. (1997). 'The applicability of project management software and advanced IT techniques in construction delays mitigation'. *International Journal of Project Management*, Vol. 15, No. 2, pp. 107–20.

COTTON, B. and OLIVER, R. (1994). *The Cyberspace Lexicon*. London: Phaidon Press.

CST (1993). *CST Version 1.01, Construction Simulation Toolkit, 3D Animation and Simulation, Users' Manual*. Jacobus Technology.

DYM, C. L. and LEVITT, R. E. (1991). *Knowledge-Based Systems in Engineering*. New York: McGraw-Hill.

EAST, W. E. and KIRBY, J. G. (1990). *A Guide to Computerised Project Scheduling*. New York: Van Nostrand Reinhold.

FISCHER, M. (1997). *4D Technologies*. Proceedings of the Global Construction IT Futures, Lake District, UK, April, pp. 86–90.

GREEN, L. and GRAY, C. (1997). 'Development construction site simulation models for concurrent engineering'. In Anumba C. and Evbuomwan, N. (Eds), *Concurrent Engineering in Construction*. London: Institution of Structural Engineering.

HALPIN, D. (1990). Simulation of site operations with CYCLONE, MicroCYCLONE Users' Manual, Division of Construction Engineering and Management, Purdue University, West Lafayette.

HARRIS, F. and MCCAFFER, R. (2001). *Modern Construction Management*, 5th edn., Oxford, Blackwell Science.

HENDRICKSON, C. and AU, T. (1989). *Project Management for Construction*. London: Prentice-Hall.

ILLINGWORTH, J. R. (2000). *Construction Methods and Planning*, 2nd edn. London: E. & F. N. Spon.

LEVINE, H. A. (1989). *Project Management Using Microcomputers*. London: McGraw-Hill

MCLELLAN, R. and MANSFIELD, N. R. (1993). 'The use of project management procedures by construction contractors'. *Proceedings*, 9th Annual ARCOM Conference, Oxford, pp. 300–10.

MAIR, G., CLARK, N., FRYER, R., HARDIMAN, R., MCGREGOR, D., RETIK, A., RETIK, N., REVIE, K. (1998). 'Integrated telepresence, virtual reality and mobile communications: a project report'. In *Mechatronics '98*. Amsterdam: Pergamon, pp. 805–10.

OLIVIERA L., WATSON I. and RETIK A. (1997). 'Intelligent visualisation as interface for CBR systems'. *Proceedings*, Mouchel Centenary Conference on Innovations in Civil and Structural Engineering, Cambridge, UK, August, pp. 311–18.

PIDD, M.(1992). *Computer Simulation in Management Science*. Chichester: Wiley.

PILCHER, R. (1992). *Principles of Construction Management*, 3rd edn. London: McGraw-Hill.

POWELL, J. (1995). 'Virtual reality and rapid prototyping for engineering'. *Proceedings*, IT Awareness Workshop, University of Salford, January.

RETIK, A., WARSZAWSKI, A. and BANAI, A. (1990). 'The use of computer graphics as a scheduling tool'. *Building and Environment*, Vol. 25, No. 2, pp. 133-42.

RETIK, A. and HAY, R. (1994). 'Visual simulation using VR'. *Proceedings*, 10th ARCOM Conference, pp. 537–46.

RETIK, A. and MARSTON, V. (1995). 'An intelligent simulation approach to cost–time forecasting for housing modernisation works'. *Computing Systems in Engineering*, Vol. 6, No. 2, pp. 177–89.

RETIK, A. (1996). 'VR system prototype for visual simulation of the construction process'. *Proceedings*, IPMA '96 World Congress on Project Management, pp. 597–605.

RETIK, A., CLARK, N., FRYER R., HARDIMAN R., McGREGOR D., MAIR G., RETIK N., REVIE K. (1997). 'Mobile hybrid virtual reality and telepresence for planning and monitoring of engineering projects'. Proceedings of 4th VRSIG UK Conference, London, November, pp. 80–89.

RIBARSKY, W. et al. (1994). 'Visualisation and analysis using VR', *IEEE Computer Graphics & Applications*, No. 1, pp. 10–12.

ROBINSON, S. (1994). *Successful Simulation*. Berkshire: McGraw Hill.

THORPE, A., BALDWIN, A., CARTER C., LEEVERS, D. and MADIGAN, D. (1995). 'Multimedia communications in construction, civil engineering'. Proceedings of ICE, 108, pp. 12–16.

TOMMELEIN, I. and ZOUEIN, P. (1993). 'Interactive dynamic layout planning'. *Journal of Construction Engineering and Management*, ASCE, 119(2), 266–87.

WHYTE, J. (1998). 'The promise and problems of implementing virtual reality in construction practice'. *Proceedings*, CIB W78 Conference, Sweden.

Chapter 6

Computer-integrated Building Representation for Design and Construction

Introduction

One of the main objectives of computer-integrated building design* and construction is to facilitate decision making when considering different aspects of the process. As far as implementation of the integration is concerned, it is essential to adopt (or even to develop) a modelling technique that will serve as a basis both for the definition of the building and for its computerised representation, from conception to completion. While it is often desirable to use a single conceptual building model throughout both the design and construction phases this is not necessarily the case with a geometric building model (Kalay, 1989). Geometric modelling techniques are adopted or developed with reference to the applications for which they are to be used. Indeed the mere term 'integration', as in 'integrated computer-aided design' or 'computer-integrated construction' (Brandon and Betts, 1995; Kumar et al., 1996; Anumba and Evbuomwan, 1997), connotes the increasing importance of a 'shared' three-dimensional (3D) model for both design and construction. However, the nature of a model that suits the typical requirements of architectural design may differ distinctly from that of a model meeting the needs of the various construction-management functions. Though many commercial CAD systems allow the mixing of several geometric representations (for lighting simulation, 3D viewing, collision testing and so on), they still do not lend themselves to capture the richness of engineering data or translate them into specific values for project variables and parameters (Kalay, 1989; Teicholz and Fischer, 1994). While many research papers imply the object-oriented technology to overcome these limitations as far as design product and construction process modelling are concerned (see, for example, 'Industry Foundation Classes' (IFC) by Wix and Leibich, 1997), this chapter proposes to improve existing modelling techniques to take advantage of the multidisciplinary construction environment.

---

*We use the term 'design' and 'designer' here in a wider sense including architectural, engineering and construction aspects.

Computer-aided building modelling: product and process representation

Representing a building (or structure) in the computer requires adaptation of the geometric modelling technique which can describe the object's physical characteristics (called 'product') to which the construction-management functions are applied and the object's behavioural characteristics – its evolution during the design process (called 'process').

Building Product Characterisation

From the viewpoint of construction management in general, and geometric modelling for construction management in particular, the object termed 'building' can be described as an assembly of a vast number of mostly non-discrete elements which are combined into one whole and together maintain a complex system of connections (for example, structural, architectural, technological, functional). The measurements of the structural elements are largely determined by the intersections between them. The same is true with regard to the areas of coatings, which are determined by the free surfaces resulting from penetrations and lines of contact between elements. Element quantities are computed by taking account of volumes 'shared' by more than one element, according to an agreed-upon set of measurement rules.

In that respect, the building is nothing like other design-based artefacts. It cannot be addressed, for example, like a machine that is composed of predefined discrete elements with fixed measurements, the quantities of which are not affected by the proximity of other elements.

Other relevant building characteristics are as follows (Shapira and Retik, 1996):

- The building is composed mainly of three kinds of elements: 3D, mostly structural elements; two-dimensional (2D) finishes of the former; and fixtures, usually measured by length or counted by pieces. The first two of these are relevant in intersections.
- Buildings are arranged into functionally defined spaces, such as rooms, halls, passages and so on. However, construction concerns may dictate a different definition, such as one that combines a number of functional spaces, or that divides one functional space into a number of smaller ones.
- Many buildings contain a certain degree of repetitiveness: within an element (such as waffle slab); of elements (identical columns); of defined spaces, along with all the elements that create them (identical hotel rooms); and even of entire storeys.

- Many building elements are always connected to certain other elements. The obvious example is doors and windows, which do not have their own 'right of existence' and are always part of the wall into which they are fixed.

Building Process Characterisation

The process modelling should allow (a) transformation (or evolution) of the object during the design, and (b) manipulation of the objects during computer-based generation of alternative solutions. Dynamic representation to support such a process, as described by Retik and Warszawski (1994) for automated design of prefabricated building, is briefly described below. It can be adopted as a possible solution in other design cases.

The product model evolves throughout the different stages of the integrated design process. The process is usually performed in several stages and by different designers. Each not only uses different computerised design tools, but also applies specific knowledge that in turn requires a different representation of the design data. At each stage one designer dominates the process, while the communication between different stages (designers) is co-ordinated through the design model by applying constraints and restrictions with the use of rules and procedures.

Dynamic representation of the model by different entities – first, a general collection of lines in layout and elevations; later, a selected set of designated supports with additional semantic information; and, finally, 3D objects with design detailing which conforms to the various design stages and also to the different design disciplines. The pattern of grouping of entities into objects can vary from one design alternative to another as a result of different parameters, constraints or rules applied at the different stages (see Figure 6.1).

The advantage of this approach is the possibility of representing the emerging design information in a meaningful form to different design professionals who usually participate in the design process (the architect, the structural engineer and the precaster (or, builder).

Furthermore, this type of representation enables easier examination of the impact of change in the various entities (both functional, such as designation of a wall as a support, or geometrical, such as changing the location of a wall) on the final design and its cost. A possibility to develop such a 'backward-trace' mechanism linking design information to both process and building models makes this approach very useful.

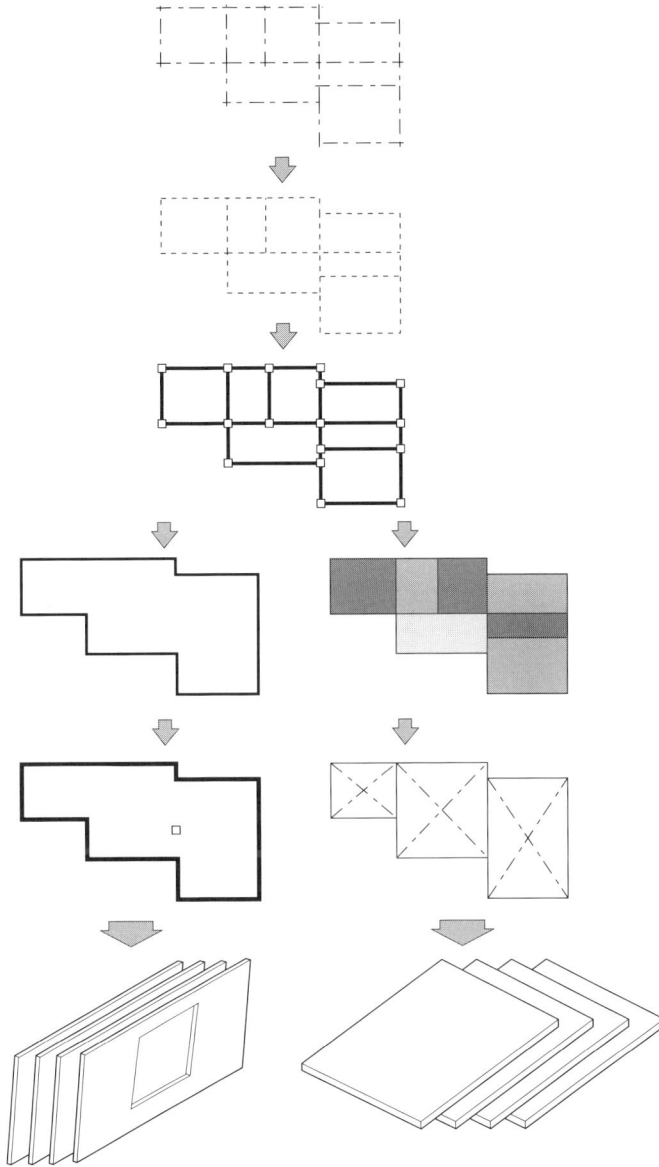

Fig. 6.1   Evolution of building representation during design

Data for Construction Management

The various construction-management functions require data of two kinds: general data and project data. ('Data' in this context also refers to 'information', that is processed data.). General data (costs, work inputs,

Table 6.1 Project information for construction management functions

Required information	Construction management functions	Examples
Element quantities by agreed classification method and measurement rules	• budgeting • owner-constructor contracting • final valuation	• quantities of a Bill of Quantities (BoQ) chapter for comparing bid proposals • overall building quantities, for the preparation of a BoQ
Quantities of elements constructed up to a certain date	• scheduling • periodic valuations • cost control	• scheduled quantity to date of a given element • actual quantity to date of a given element
Element quantities in predefined building zones	• scheduling • budgeting	• quantities of main items in a modular 4-apartment section • quantities of foundations (i.e., excavation, rebar, concrete), for comparing foundation alternatives
Element quantities by different classification method and/or measurement rules	• subcontracting	• overall quantity of different-width masonry walls plastering quantity, without deducting area of openings
Quantities of elements by a running number	• periodic valuations	• concrete in columns from column number Y to column number Y
Overall material quantities in the building	• site production planning • subcontracting • cost estimating • material management	• overall quantities in the building, for ordering concrete supplies • overall quantity of floor tiles for planning transportation to the site

Table 6.1 *continued*

Material quantities in predefined building zones	• site production planning • cost estimating • material management • daily operation management	• concrete-pumping: quantity of a group of elements for ordering concrete and planning the pour • quantity of tiles by storeys, when interim site storage is not available
Quantities in vertical building segments	• site production planning	• quantities of all concrete elements covered by a tower-cranes' boom
Element quantities by rooms	• budgeting • periodic valuations • daily operation management • maintenance management	• flooring and painting quantities by rooms, for submission of progress reports • quantity of ceramic tiles in kitchen, for finishing alternatives comparison
Overall building geometry (shape and measurements)	• site production planning	• examination of conveying equipment (crane, concrete pump) • siting of concrete plant
Location of elements in the building (by storey, wing, room, etc.)	• scheduling • site production planning • cost estimation	• storey (high/low) in which light partitions are to be erected • location of heavy steel door, for planning its conveying and assembly

material inputs, equipment technical data) are not derived from the building plans and are not necessarily specific to a particular project. Project data are, for the most part, derived from the building plans (and possibly other project documents), and are the kind of data that the geometric building model (along with its non-geometric attributes) is expected to provide. These project data are essentially quantitative, and are inseparably connected with the building geometry and the location of the elements in the building space. The project data are expressed in quantities of building elements sorted out (grouped and separated) by various keys: in measurements of elements and distances between elements; in repetitiveness of elements and their degree of complexity; and in other defining characteristics. These data are needed not only for the classic planning, management, and control functions (such as cost estimating, scheduling, site-production planning, materials management), but also for other functions that comprise a broader definition of 'construction management' (such as detection of design errors and compatibility of plans, comparison of design alternatives, maintenance management and the like).

Table 6.1 details typical building data and information required for construction management. These were identified in a study of various construction-management functions, within the search for the most appropriate geometric building model (Shapira and Retik, 1996).

Representation of Building-Element Relationships

For a common denominator, especially in light of the building characteristics, the management functions look to elements – quantities of different levels of detail and of different subdivisions, sections defined in the structure, and so on – all derived in one form or another from spatial relationships ('encounters', or, 'intersections') between elements. Thus, intersections stand out as the central issue, and from this follows the need to focus on them when first examining the suitability of different modelling techniques. The intersections – penetrations and contacts – between the elements of a building are characterised by a number of properties:

- Every building contains large numbers of intersections: columns connect with the floor beneath them and with the ceiling/beams above them, and sometimes also with masonry walls; walls are involved in similar intersections and also intersect one another. The number of intersections is in the same order of magnitude as the number of elements in the building's skeleton.
- Some intersections are complex and involve more than two elements, such as two beams encountering a column at the point of their intersection.

- Many and variegated geometries – some very complex – are involved in the intersections. The geometries of intersecting elements should be addressed, as well as those of the resulting forms.
- The construction-management perspective of these intersections is different from that of other aspects (architectural design, functional assessment, and such like) and demands adherence to a rigid element-classification system and a set of measurement rules.

These properties, individually and collectively, exert a decisive influence on the selection of the geometric modelling technique. Dominant among them, however, is the number of intersections. This importance stems from the direct effect this characteristic has on the problem of locating the intersections; examining the elements' juxtaposition in space with a view to determining which element connects with which other element(s) and where such connections occur. The other three characteristics mentioned are all related more to the problem of calculating the intersections, such as determining the type of intersection (point, line, surface contact or penetration) and its geometric properties. Their effects are described by Shapira (1993) and Shapira and Retik (1996).

References

ANUMBA C. and EVBUOMWAN N. (Eds) (1997). 'Concurrent Engineering in Construction'. Proceedings of International Conference on Computing in Structural Engineering, London: SETO.

BRANDON P. and BETTS, M. (Eds) (1995). *Integrated Construction Information*. London: SPON.

KALAY, Y. E. (1989). *Modelling Objects and Environments*. New York: John Wiley.

KUMAR, B., MACLEOD, I. and RETIK, A. (Eds) (1996). 'IT in Civil & Structural Engineering Design'. Proceedings of ITCSED '96, Inverleith Spottiswoode, Glasgow.

RETIK, A. and WARSZAWSKI, A. (1994) 'Automated design of prefabricated building', *Building and Environment*, Vol. 29, No. 4, pp. 421–36.

SHAPIRA, A. (1993). 'Octree subdivision of building elements'. *Journal of Computing in Civil Engineering*, ASCE, 7(4), pp. 437–59.

SHAPIRA, A. and RETIK, A. (1996). 'Computer-integrated construction management: a case for octree-based representation'. *International Journal of Construction Information Technology*, Vol. 4, No. 1, pp. 61–82.

TEICHOLZ, P. and FISCHER, M. (1994). 'Strategy for computer integrated construction technology'. *Journal of Construction Engineering Management*, ASCE, 120(1), pp. 117–31.

WIX, J. and LEIBICH, T. (1997). 'Industry foundation classes: architecture and development guidelines'. In Drogemuller, J. (Ed.), *Proceedings of the CIB W78 on IT Support for Construction Process Reengineering*, Cairns, Australia, July 1997.

Chapter 7

Computer-integrated Multidisciplinary Concurrent Engineering

'There are many who can tell what is wrong with a design when it has been completed but almost no one can say exactly what is needed at the outset.'

Professor Medland, Bath University

Introduction

Traditionally (and logically) architectural design preceded construction in utilising computerised building models. As a result, the realisation of a building starts with design of varying levels of completeness and only then is followed by construction, which may be done in sections as the design develops. It is no wonder, therefore, that integrated systems conceived to support both design and construction management have traditionally been focused on representation schemes that better meet the needs of the former and not necessarily those of the latter (see Chapter 6).

Efforts to shorten building delivery time (both design and construction), with the same or possibly enhanced quality, have introduced the idea of concurrent engineering, which aspires to integrate different designers and engineers in an integrated multidisciplinary process. This approach has found wide application in the manufacturing industry (Prasad, 1996; 1997), which has already provided the construction industry with some useful and successfully adopted techniques. The relevance of concurrent engineering to the construction industry is clear. Moreover, novel procurement processes, such as design–build, build–operate–(own)–transfer (BOT/BOOT), have created an encouraging ground for adaptation of the concurrent engineering approach in construction (Anumba and Evbuomwan, 1997; Fenves et al., 1993; Kumar et al, 1996). However, the fragmented nature of the construction industry implies that not only are design and construction very often separated and carried out by different parties, but the designers are also from different organisations possibly in different locations. This approach warrants some sort of adaptation. The main problem concerns the communication of design information and co-ordination between different design parties. Unless otherwise stated, 'design' applies to all stages from preliminary design to production planning; 'planning' can also be defined as a process design.

As far as computerisation (or automation) of the process is concerned, this problem may be approached from two different angles. At a lower level, the more pragmatic approach is to solve the problem of data and information exchange concentrating primarily on the standardisation issues. For example, a designer may produce drawings using CAD software and pass them onto another designer on a disk or via computer networks. EDICON (Electronic Data Interchange Construction), EDM (Electronic Data Management) and CITE (Construction Industry Trading Electronically) are examples of UK and US initiatives in this area (there is more on these in Baldwin et al., 1996). The limitation of this approach is that each designer works in isolation and the interactions between the different designers' activities are not taken care of.

The higher level approach is not only to provide for access and exchange of the design data and information but also for the co-ordination between the multidisciplinary design participants for consistent management of the design process. Thus, the influence of one designer's changes to the design is automatically reflected in the other participants' designs. Although considerably more complex than the previous approach, this approach attempts to manage the design process much more effectively.

Many research and development projects concerned with computerisation tackle the automation, modelling, representation, co-ordination and even re-engineering of design and construction processes with different objectives (see, for examples, Anumba and Evbuomwan, 1997; Drogemuller, 1997; Kumar and Retik, 1996; Port, 1993). Though a wide range of approaches exists, the application of information technology tools can be grouped in the context of this chapter into two main themes:

- the development of integrated-design processes by sharing data and information between different stages
- the development of advanced tools for different stages of the construction process (such as CAD packages, planning packages).

Under both themes the problem of the information exchange between different stages and packages has arisen and has led to attempts at the creation of integrated databases or common data banks, which should avoid double or triple sets of data. The different proposed solutions still have problems, not only in how to derive appropriate data from these banks but also how to analyse the retrieved data.

Preliminary Design Stage

At the very beginning of the design process, in terms of computer-aided design (CAD), the main problem is linking the design data and information which exist in various data banks (structural elements, specifications, cost of materials) and processes (productivity of work teams, technological solutions) with preliminary design drawings where the architect's solution is described in terms of spaces represented by lines, surfaces, symbols and so on.

In addition, computerised drawing is, in the majority of cases, the source of data acquisition for a system. Moreover, the influence of decisions at the initial design stage makes it very important to provide the designer (architect/ structural engineer) with appropriate tools to evaluate several alternatives at this stage. Therefore, improved access to design information at the preliminary design stage by providing linkage between drawings and integrated databases will create a tool for decision making by evaluation, for example, of the cost, time and constructability of different solutions.

There are obvious difficulties in producing a set of unified evaluation criteria to embody all considerations, such as aesthetics, stability and cost. However, it is very important for design-analysis purposes to trace their interrelationship and their influence on a particular design solution.

Different considerations of design represent functions and goals of the various participants of the design process. Architectural considerations generally shape the decision-making process. The purpose of structural considerations is to provide the required strength and stability to the building in order to conform to different structural codes of practice. Economical consideration is an indicator (sometimes decisive) in the evaluation of different solutions. However, the cheapest solution is not always chosen. Sometimes a unique architectural solution increases the value of the building beyond the additional cost of its production.

The communication between different designers may be established by the notional link between the functional elements of the design model. References to the objects are possible according to their geometry, topology or semantic attributes. Due to the different geometrical interpretations by the architect, structural engineer or quantity surveyor of a line section with a semantic attribute such as 'exterior wall', it is possible to create appropriate geometry of the object for every designer, relating to the same line sections. The topology rules (mainly derived from architectural and structural considerations) can be compiled together with the geometrical rules into spatial grammars (Stiny, 1980 and 1991), dealing with the feasibility and constructability of a particular solution. The issue of misinterpretation also needs to be addressed.

Multidisciplinary Representation of the Planning and Design Process

The previous chapter described the concept of dynamically linking the representations of evolving design models. Such a link allows 'vertical' integration of the design process. The approach presented here is intended to support 'horizontal' (concurrent) integration on every vertical level (design stage). This approach was postulated by, and a prototype developed by Retik and Kumar (1996).

The idea behind the approach is to represent the building model in different ways compatible with the representation of the knowledge to be applied to this model at each stage of the planning and design process. The main thrust of the approach is to develop a flexible representation of design information. The ultimate goal is to be able to interpret this representation intelligently. To this end, it is important that both interpretation and representation go hand in hand. However, the intelligent interpretation is the long-term goal; in the short term we are looking at providing appropriate information to designers in a more effective way. The goals in the short term will be limited to responses to intelligent queries from designers. The information presented to the designers should be as transparent as possible so that they can immediately see what they are looking for. This will be the main driving force behind our strategy for the development of a representation of design information.

In short, it is important to provide a multidisciplinary representation of design information that makes available as much relevant information to a designer as possible, by making explicit most of the 'hidden' information in a design drawing. The automated interpretation of this information is our long-term, ultimate goal.

Computer-integrated planning and design

The conceptual framework of the proposed prototype is illustrated in Figure 7.1.

CAD Drafting and Data Acquisition

The 2D drawing technique is used in the majority of CAD packages as an alternative to drafting-board systems. These 2D CAD facilities not only make the production of drawings as a design output easier, but also provide a convenient tool for the creation of several alternatives at earlier design stages.

The majority of the CAD systems present drawings in the form of points, straight lines, circles and arcs of varying thickness and colour; they also allow textual information to be attached to these drawing entities. Although a drawing is still being used as the main communication medium between different designers, the main disadvantage of such a 2D representation is that other designers have to interpret the 3D objects from the 2D drawings. Fortunately, this task is usually not as complicated as in mechanical engineering cases (intersection between a sphere and cone, for example), but it is still a source of mistakes and misinterpretations.

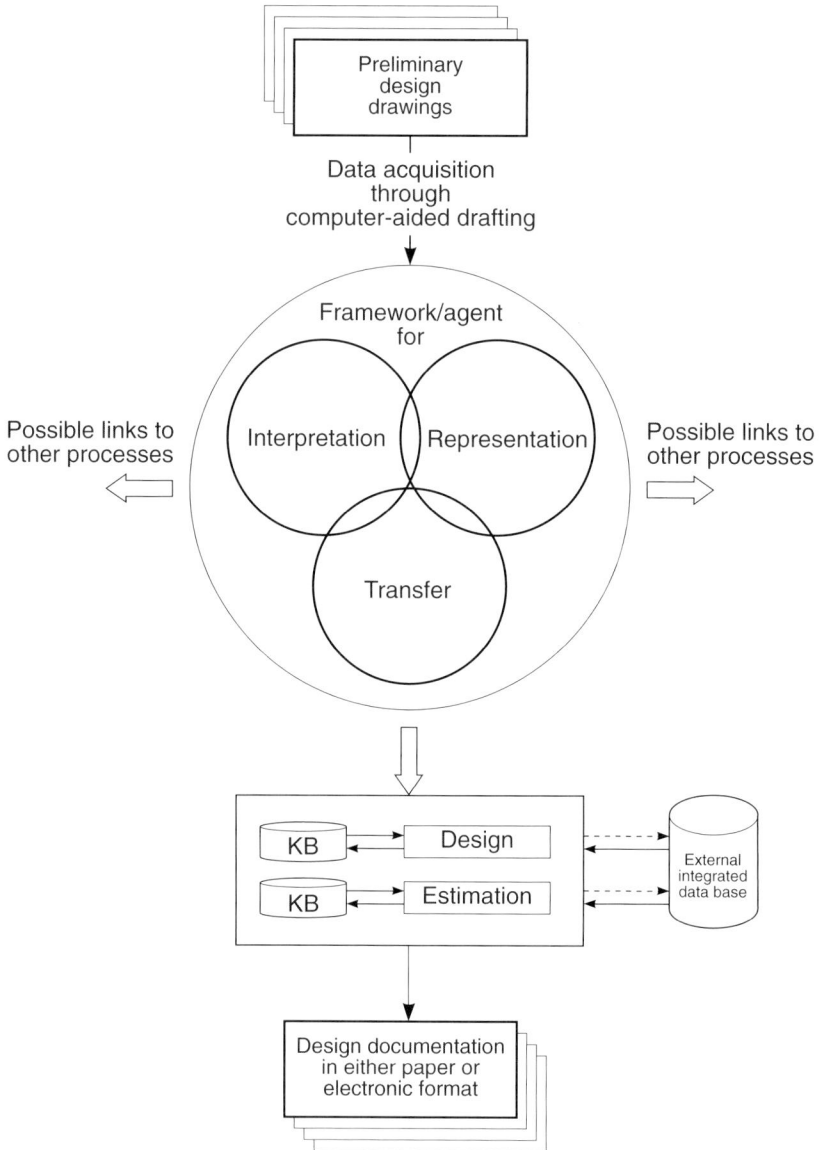

Fig. 7.1   Conceptual framework for multidisciplinary design process

On the other hand, it is possible today to design a building using 3D modelling. Though there are many benefits in some cases (space arrangements; clash detection, i.e. automated search and prevention of conflicts in designs of, for example, pipe collisions, walls/holes overlaps, etc.; and so on), there are drawbacks in others. For example, the amount of visible and invisible data (such as services, reinforcement, finish details – see Retik, 1995) is such that it can completely blur the graphical presentation of the design. This requires additional effort during design to allow for efficient presentation and communication of the design models. So, 2D or not 2D is (still) the question! However, the approach presented below for 2D design presentation and interpretation could be applied to 3D models or, more likely, a mixture of 2D and 3D representations.

Representation and Interpretation of Drawing Information

When CAD is being used, it automatically creates a database which contains records of drawing entities. A drawing entity is a basic shape or form which is used in a drawing, such as point, line, arc, polyline and even text. Every entity (a line, for example) has at least graphical (line type, colour and such like) and geometrical (co-ordinates of end points) data presented as a record in the database. The user may provide additional verbal (semantic) information called attributes, which should be linked to an entity or a group of entities.

The system data can be drawn even from the conceptual design draft layout. This information is entered through a 2D representation of the drawings (by using a CAD package). Context data, representing factual information on the particular design layout, will be obtained by means of a special program which translates the computerised drawing (ASCII format) into a collection of basic graphical entities (ordinary lines, multiple lines and so on). The properties needed to be defined for each entity are geometrical co-ordinates, colour and type of line, special attributes and such like. It is necessary to sift the relevant data and organise it in an appropriate structure (called a 'template') for convenient access, thereby eliminating the influence of the personal drawing technique. In addition, the dimensions of, and distance between, the lines can be adapted (if necessary) to a modular grid for cases using modular components, or to specific regulations. The program that performs all these activities will interface between the system and external CAD. At this stage, the design information will be organised in architect-oriented functional elements such as walls and partitions; the purpose of which is to divide the design space into functional spaces.

Information Transfer

Most CAD systems provide facilities to export information to non-CAD systems such as database systems (dBASE file format for example). But, not all information can be transferred, mainly because of different representation and product models. Moreover, implicit information (such as the topology of drawing elements) may be lost if it is not considered properly.

Creating a link between CAD design and an external commercial system is a complicated task. The key problem here is that not only are the external systems (such as structural-analysis packages) or databanks (such as cost information) task oriented and made independent of specific drawings, but they also use different functional elements which conform with professional practice and knowledge as well as the physical products involved. So, if we are interested in accessing the existing databases or communicating with other design partners, the design data represented by architectural elements (spaces, areas) should be interpreted in structural elements (beams, bearing walls) and managerial elements (costs, activities). The important thing in such interpretation is also to define dependencies between such elements, so that interdisciplinary analysis is possible.

Structural Design

The purpose of the interface in relation to structural design is to provide information regarding structural elements required in a building. For example, a line can represent a wall or a beam or a number of walls or beams. These walls could simply be partition walls or load-bearing walls. When a structural engineer looks at a drawing, he or she can easily infer which ones need to be structural elements and which ones could be non-structural. It is not a trivial task to make the computer do the same. However, well before the structural elements have to be identified, the overall structural pattern has to be decided in order to choose appropriate structural systems.

One possible solution for identifying structural and non-structural elements may be attaching labels to the geometrical entities such as lines and arcs. However, this strategy fails when more than one element being represented by one single geometrical entity, which is quite often the case.

Another approach could be to use case-based reasoning. Case-based reasoning is an approach to solving problems based on past solutions to similar problems. A case base of past designs which include architectural plans along with the corresponding structural layouts could be maintained and used in future designs. The case base could be searched for similar plans to the one currently being designed and the corresponding structural layout retrieved for the purposes of structural design.

Of course, a number of issues are involved in such an approach. The most important of all is finding the similarity between past and present plans. It is quite likely that an exact match is not found, in which case similarities can be based on building up some sort of similarity matrix. Although this approach might not give a ready-made solution, one can at least find a starting point which can be adapted appropriately for the problem at hand. The ultimate objective of such an approach would be an automatic adaptation. An approach to case-based design and its application to structural design can be seen in more detail in Kumar et al. (1997). This approach is based on utilising methods used in the past to solve similar problems in order to arrive at patterns, rather than using patterns for similar problems from the past. In that sense, the approach is different from more conventional case-based design approaches which concentrate on using similar designs rather than design methods; it therefore deserves further discussion.

Method-Based Representation

The motivation behind this work is the observation that, even though structural-engineering design, in general, does not follow a fixed procedure, the individual designs have a fixed procedure, involving qualitative and quantitative analysis, interpretations, heuristics and design decisions. The sequence of steps followed to arrive at a design is called a method. In this method of design, there is a case base of methods rather than a case base of solutions.

In this approach, design methods are represented in the form of a decomposition involving a hierarchy. The method for a particular task has several subtasks which have their methods and so on. A method is a data structure used to capture design knowledge in the form of operations and rules, along with their sequence.

The decomposition of tasks, rather than the solutions, avoids several problems encountered in the conventional approaches. Even when solutions are not decomposable because of the strong interaction between their parts, tasks are decomposable because of the sequential nature of human problem solving (Simon, 1992).

Contents of a case

A case contains the following components:

- a set of preconditions for which the design is generated
- the definition of the design task
- the method used to arrive at the design.

The preconditions are represented as facts, constraints or variable-values. The definition of the design task is also represented as a fact.

Indexing and retrieval of cases

Indexing is based on the trigger state, which refers to the conditions under which a method could be invoked. If there is no operating pattern defined within the trigger-state, this indexing scheme results in the selection of all the methods having the same objective. During the execution of a method, all the similar methods for the subtasks are retrieved from the respective groups.

Managerial Information

The details which managerial-information designers are likely to be interested in at the preliminary design stage are cost and time estimates. Approximate estimates (such as base unit price of square-metre-of-floor-area) are not good enough for analysing several design alternatives (which, for example, may have the same floor area). On the other hand, there are several elemental-analysis estimating techniques (Ahuja and Campbell, 1988; Ferry and Brandon, 1991) which use standard cost elements and are supported by well established information sources such as the Building Cost Information Service in the UK and Means and Boeckh in the USA. In the system the information may be acquired and stored with the aid of a database management system. The database will contain all information needed for cost estimation according to the cost elements. To retrieve actual elements the knowledge for the interpretation of architectural and structural elements into the cost elements should be integrated. This knowledge is likely to be organised in different modules according to the different building types; or if an external knoweldge-based system is available it may be integrated instead (as demonstrated by Retik and Warszawski, 1994).

Production Planning

Once the necessary information on a project is available the planner begins with master-plan generation. At this initial planning stage the design product (which is actually presented in its final 'ready-to-use' or 'turnkey' shape) should be broken down into construction components; a sequence of activities to implement the work established; appropriate resources identified; and plant and equipment selected. This task is a heuristic generation of a construction process. It is based not only on experience but on the ability to generate 3D images from 2D design drawings. Very few alternatives are considered at this stage because this would be time consuming. After the feasibility of the chosen process has been verified, activities can be set into a time schedule, taking into account existing constraints, necessary milestones and possible resource optimisation. The time schedule is a key tool for monitoring and control of the project progress at the construction stage. In addition to activities directly associated with construction of the design product there are auxiliary (or preliminary) activities mainly related to the equipment and organisation of the construction site. These activities must also be monitored.

System Development

The described design process is performed in several stages by different designers. Each uses not only different design tools, but also applies specific knowledge which in turn requires a different representation of the design data.

To deal with these, the following strategy is adopted for system development:

- task-oriented modularity of the system structure
- independence of different parts and modules
- integration of different existing packages to enhance the power of data processing.

As a result, it is possible, first, to control the design process by establishing the order of the task modules' firing (the function of an inference engine); second, to simplify the development of the system; and, third, to simplify the access and update of different modules.

The subdivision of the knowledge bases into modules is particularly important for easier development. It also allows integration of design-specific packages and facilitates the updating and substitution of technological solutions.

References

AHUJA, H. N. and CAMPBELL, W. J. (1988). *Estimating: From Concept to Completion*. New Jersey: Prentice-Hall.

ANUMBA C. and EVBUOMWAN N. (Eds) (1997). *Concurrent Engineering in Construction: Proceedings of International Conference on Computing in Structural Engineering*. London: SETO.

BALDWIN, A., THORPE, A. and CARTER, C. (1996). 'The construction alliance and electronic information exchange: a symbolic relationship'. In Langford, D. A. and Retik, A. (Eds), *The Organisation and Management of Construction*, Vol. 3. London: SPON, pp. 23–32.

DROGEMULLER, J. (Ed.) (1997). *IT Support for Construction Process Re-engineering*. Proceedings of CIB W78 Working Commission on IT in Construction, Cairns, Queensland, Australia, July.

FENVES, S., FLEMMING, U., HENDRICKSON, C., MAHER, M. L., QUADREL, R., TERK, M. and WOODBURY, R. (1993). *Computer Integrated Concurrent Building Design*. New York: Academic Press.

FERRY, D. J. and BRANDON, P. S. (1991). *Cost Planning of Building*. Oxford: BSP Professional Books.

KUMAR, B., RAPHAEL, B. and MACLEOD, I. (1997). 'CADREM: Case-based system for structural design'. *International Journal of Engineering with Computers* London: Spinger-Verlag. Vol. 13, pp. 153–64.

KUMAR B., MACLEOD, I. and RETIK, A. (Eds) (1996). *IT in Civil and Structural Engineering Design*. Proceedings of ITCSED'96, Inverleith Spottiswoode, Glasgow.

KUMAR, B. and RETIK, A. (Eds) (1996). *Information Representation and Delivery*. Proceedings of ITSCED'96. Edinburgh: CIVIL-COMP Press.

PORT, S. (1993). *CAD Management*. London: Blackwell Science.

PRASAD, B. (1996) *Concurrent Engineering Fundamentals. Vol. I: Integrated Product and Process Organisation.* New Jersey: Prentice Hall.

PRASAD, B. (1997). *Concurrent Engineering Fundamentals. Vol. II: Integrated Product Development.* New Jersey: Prentice Hall.

RETIK, A. (1995). 'Discussion on geometric-based reasoning system for project planning'. *Journal of Computing in Civil Engineering* , October, pp. 293–4.

RETIK, A. and KUMAR, B. (1996). 'Computer-aided integration of multidisciplinary design Information'. *Advances in Engineering Software*, 25, pp. 111–22.

RETIK, A. and WARSZAWSKI, A. (1994). 'Automated design of prefabricated building'. *Building and Environment*, Vol. 29, No. 4, pp. 421–36.

SIMON H. A. (1992). 'People and computers: their roles in creative design'. Keynote address, AID'92, CMU, Pittsburgh, USA.

STINY, G. (1980). 'Introduction to shape and shape grammars'. *Environment and Planning* B, Vol. 7, pp. 343–51.

STINY, G. (1991). 'The algebras of design'. *Research in Engineering Design*, Vol. 2, No. 3, pp. 171–81.

Part 3

Applications of IT in the Construction
Business

Chapter 8

The Strategic Use of Information Technology

Introduction

In the postwar period the use of technology has been vaunted by politicians and business leaders, and approached with caution by trade unionists and environmentalists. In this chapter we will be looking at how technology can be used to help provide competitive advantage to construction companies.

The Current Use of Information Technology

Information technology (IT) has long been central to the processes used by the construction industry in the following ways:

- Developments in computer-aided design and drafting (CADD) have helped to automate the design process and reduce the many laborious hours spent by draftspersons in drawing up production information for use on site.
- Design engineers use computer software to help them quickly and accurately carry out calculations.
- Most contractors will use project planning software.
- The accounts office will use automated payroll, ledger and forecasting systems and will make great use of spreadsheets.

In short, computer applications can be found in design, engineering, management, building economics and administration. Such developments are often ill co-ordinated and underused and so users have few opportunities to harness the current IT capabilities. This problem was recognised by Brandon et al. (1989) who sought to consolidate and integrate these elements into a composite expert system to advise clients on the strategic planning of particular construction projects. Clients are advised of the economic feasibility, the economics of building geometry and the consequences of various procurement strategies. This vision can assist IT to be used in the arena of strategic management and away from applications in particular operations such as accounts or project planning.

Gallacher (1988) detected three stages in the development of IT applications (Table 8.1) and the trend is clearly towards greater sophistication and wider use of IT in strategic management applications. Many firms operating within the construction industry will remain and choose to remain in Gallacher's Era I, while earlier adopters of technology will be seen to inhabit Era III. This is a microcosm of the varied nature of the construction industry which comprises many different firms offering different services on a very wide range of projects.

This heterogeneity will contain firms which are driven by information and those involved in the physical work involved in making a building grow. A bricklayer will want to know where to build a wall, with which bricks and which bond, to what height, and where any openings occur. The presentation of the product is then dependant upon his or her skill in their chosen craft. The information content of the design process in construction is high; so much so that the British construction industry has spent years in establishing methods by which construction data would be co-ordinated and subsequently be based upon a computer-based transaction system. Porter and Millar (1985) model the differences between the informational content of processes and product (Figure 8.1).

The location of the construction industry into the two sectors of the quadrant in Figure 8.1 is obviously stereotyping firms operating in these areas of the construction industry. Movement across the sectors is possible, and even desirable, to gain competitive advantage. For example, the use of a back-hoe excavator to dig a trench may require little product information (the length, depth and width of the trench), but the same machine presented in a robotic form to work in, say, the nuclear waste industry has changed to being high in process information and product information. The change from a carbon-based (human) operator to a silicone-based (robotic) operator has changed the informational environment. (Such comparisons are not fanciful,

Table 8.1   Stages in IT applications

Characteristic of era	IT era		
	I (1960s)	II (1970s & 80s)	III (Late 1980s +)
Objective	Transaction processing	Management decision support	Competitive advantage
Primary target	Clerical/ administration	Individual managers	Complete layers of business
Business justification	Productivity and efficiency	Effectiveness	Strategy

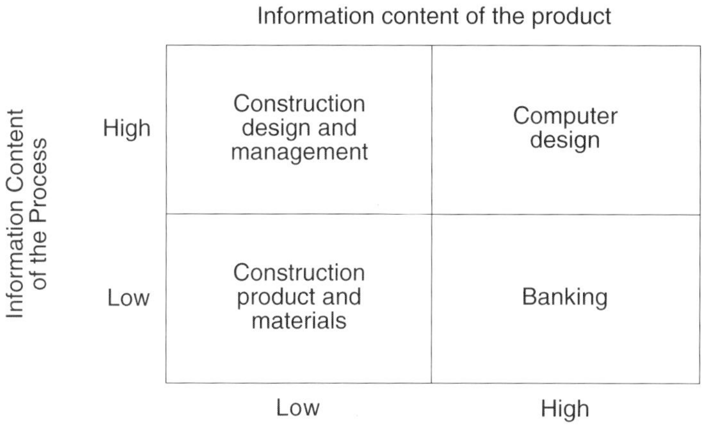

Information content of the product

		Low	High
Information Content of the Process	High	Construction design and management	Computer design
	Low	Construction product and materials	Banking

Fig. 8.1 Information contents of products and processes

for a robotically controlled excavator is almost ready for market.) However, this illustration typifies the use of IT in construction – it has been used for specific applications and it may be anticipated that the future wave will be for IT to be used for strategic purposes rather than merely seeking efficiency gains.

Earl (1989) maps the growing level of sophistication in the use of IT in a strategic context (Table 8.2). In Betts et al.'s (1991) terms, some firms and sectors are 'delayed' while others have reached the 'drive' stage of development, at which point they are using IT in a strategic way.

Betts et al. (1991) envision a much wider use of IT for strategic management. They evaluate the current use of IT in construction and seek to specify where IT can be used within a framework of three different strategic management concepts. Two of the frameworks are drawn from Porter and Millar's (1985) work on competitive analysis (Figure 8.2); the third is drawn from Flaaten et al. (1989) and proposes the use of value chains to achieve competitive advantage. Betts et al., using Porter's model, gauge the current position in respect of IT use for strategic management and present what they see as opportunities for future use at different levels in the industry. Betts et al. use the four outer boxes of Porter's view and include the turbulence created at the centre as a fifth dimension which they call 'jockeying for position' (Table 8.3). Figure 8.3 illustrates generic strategies which may be used to gain competitive advantage.

One may challenge Betts et al.'s interpretation of the current use of IT and its prospective use, but the framework for analysis is undeniably powerful. If nothing else it maps out the needs for the strategic use IT and this can move the debate away from mechanisms of automating existing processes to a more challenging model of what needs to be done to gain competitive advantage from IT.

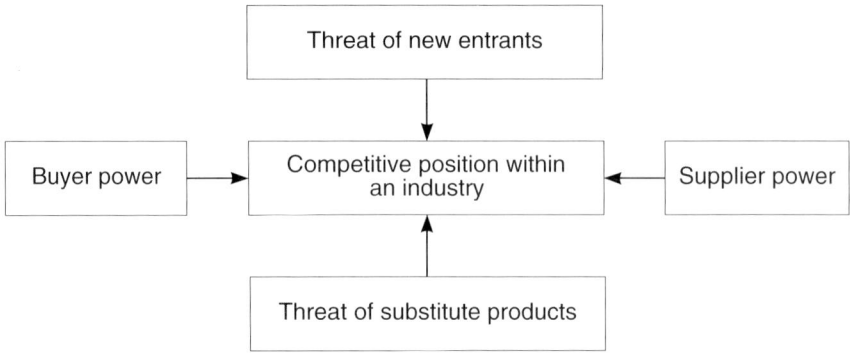

Fig. 8.2 *Competitive forces*

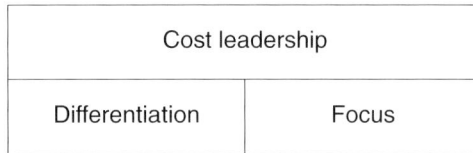

Fig. 8.3 *Porter's generic strategies*

Table 8.2 *IT in a strategic context*

Strategic context	Characteristic	Metaphor
IT has no strategic impact in the sector	Opportunities or threats from IT are not yet apparent or perceived	Delayed
IT is the means of delivering goods and services in the sector	Computer based transactions underpin business operations	Delivery
Business strategies increasingly depend on IT for their implementation	Business and functional strategies require a major automation; information or communications capability and/or are made possible by these technologies	Dependent
IT potentially provides new strategic opportunities	Specific applications or technologies are exploited for developing business and changing ways of managing	Drive

Table 8.3 Strategic opportunities

Part of strategic approach		National	Professional	Enterprise	Project	Building (component)
Value chains (Flaaten)	Lower cost	1	2	2	2	2
	Higher value		2	2	2	2
	Value channels	1	1	1	2	1
Five forces model (Porter)	New entrant		2	1		
	Supplier power		1	1	2	
	Substitute product		1	1		1
	Jockeying for position		1	1		
Generic strategies	Product differentiation	2	1	2	1	2
	Cost leadership	1	2	2	1	2
	Product focus	1	1	1	1	2

Note: 2 represents a greater strategic opportunity than 1.

Competitive Advantage and IT

The Construction Industry Computing Association (CICA, 1990) envisage that IT will play an increasing part in sustaining competitive advantage in construction organisations. One may argue that this has been a repeated call since the 1970s, but the way IT has been used to serve business ends has been highly fragmented in the construction industry. Badly co-ordinated investment policies have left construction organisations, in the main, with isolated and uncoordinated IT applications. Mraovic's (1996) research on five organisations in Croatia found that the use of IT is suppressed by uncontrolled and unsynchronised implementation of computer systems, leading to unfulfilled potential. Moreover, in these companies IT had not been used in support of the firm's strategic goals; rather, the 'techies' were working independently of the strategic managers of the companies. In short, IT was not used as an instrument in support of strategic goals.

Mraovic (1996) argues that success is likely only if corporate cultures are harnessed to ensure that information is seen as a corporate resource. This will require an integration within IT systems, and between IT systems and the strategic plan for an organisation. This phenomena is not only evidenced in contractor organisations: the Royal Institute of British Architects (RIBA, 1992) note that large architecture practices have difficulty in effective management of IT investments.

These observations pose questions concerning a managerial strategy for using IT for strategic management. This question can be inverted to note that a strategic management is a vital prerequisite to a successful IT strategy (James Martin Associates, 1988). Strategic management according to Ansoff (1985) is a 'systematic approach to a major and increasingly important responsibility of general management; to position and relate the firm to its environment in a way which will ensure its continued success and make it secure from surprises'.

Ansoff's definitions highlight several key issues that have a bearing on the application of IT. First, Ansoff identifies strategic management as 'systematic'. This implies a view ahead is being taken: envisioning the opportunities that are coming up, anticipating downturns in a particular sector or shifts in interest rates, etc.; in Ansoff's terms, protecting the firm against surprises. This proactivity (though not always observed in practice), where managers may react to events rather than plan for them, depends heavily on an information system operating within the firm which is managed by appropriate information technology. The information system can assist the general managers of a construction firm 'to position and relate the firm to its environment'; positioning the firm within the various construction markets will be an essential component of strategic management – which services to offer (construction management, management contracting, project management, etc.), which markets, (industrial commercial, speculative housing, etc.), in which geographical area should the firm work, will all be part of the positioning process. Information will be part of the analysis that will help to position the firm. It will also assist in identifying aspects of the environment that will need to be scanned if the position of the firm in the market is to link with the environment. Again, information will be essential in conducting this environmental analysis. Alshawi and Aouad (1993) illustrate diagrammatically (see Figure 8.4) how information technology is linked to strategic management. An early definition of the business objectives will create a strategy for a firm, and to be successful these objectives need to be quantified and criteria for their achievement identified.

Alshawi and Aouad's model links the IT function to strategy. In practice competitive advantage requires technology to be used in a different or more productive way. In construction obvious examples may be seen at head office in office automation, computerised purchasing, materials management, design, estimating and surveying functions. On site a computer-based planning function is a ubiquitous part of the technology, which may be used to lower cost, or differentiate a firms service to clients.

Hampson (1996) sees the application of technology for competitive advantage in a rather broader way. In a study of a Californian road builder he investigated how investments in various technologies (including IT) affected the competitive performance in different industry sectors. A firm's technology strategy needs to match the characteristics of the sector of the construction market in which the firm is seeking to position itself. Strategic fit between the technology strategy and the business environment is seen by Hampson as

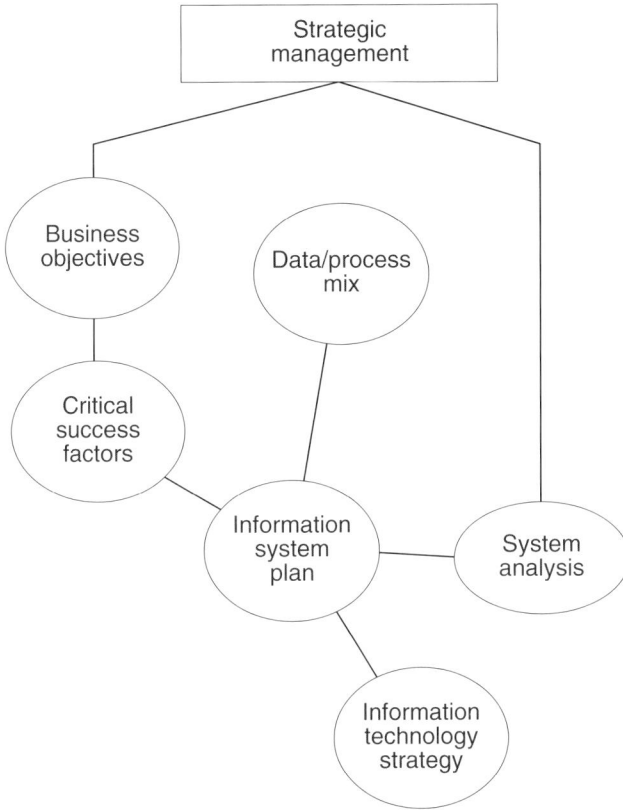

Fig. 8.4   Diagrammatic representation of the link between IT and Strategic Management

highly desirable. The technology strategy comprises five key conceptual dimensions (as shown in Table 8.4) and each of these concepts contained a number of elements.

Each of Hampson's elements were matched against three niche markets in the road construction industry, namely:

- minor works, maintenance, modification to existing roads
- new works which are routine
- special projects which are large and complex in scope.

The technology requirements for each of these submarkets were appraised for each element, as shown in Table 8.5, and then collapsed into summary of the technology requirements in order for companies to offer superior technology in each of the submarkets. Clearly much of the 'technology' to which Hampson refers will be areas other than IT; however, the same methodology can be applied to assessing the IT requirements to position a firm into specific submarkets.

Table 8.4   Strategic fit and technology strategy

Technology strategy measures	Industry niche		
	A. Minor modifications	B. Routine alignment	C. Special projects
(a) Competitive positioning	M	H	H
Emphasis of technology in overall business strategy	M	H	H
Command of key technologies in sector	H	H	H
Command of unique technological position	M	H	H
Ability to be key technology leader	M	H	H
Monitoring of competitor technologies	M	H	H
(b) Sourcing of technology	M	H	H
Acquisition of explicit technology	H	H	H
Acquisition of implicit technology – head office management	H	H	H
Acquisition of implicit technology – site management	M–H	H	M–H
Emphasis on organisational learning	M	H	H
Monitoring evolving technologies in sector	M	H	H
(c) Scope of technology strategy	M	L	H
Breadth of technological capabilities	L	L	H
Content focus of tech monitoring and development	M	M	H
Geographical focus of technological monitoring and development	M	L	H
(d) Depth of technology strategy	L–M	M–H	H
Emphasis on research and development	M	H	H
Depth of technology capabilities – head office management	M	H	H
Depth of technology capabilities – site management	M	L–M	H
Degree of specialist tasking	L	H	M
(e) Organisational fit	L–M	M–H	H
Reward systems – head office management	M	M–H	M–H
Reward systems – site management	M	M–H	M–H
Structuring of information flows – site to site	L–M	M–H	H
Structuring of information flows – site and head office	L	H	H

Key: H – High; M – medium; L – low

Table 8.5   Technology requirement in construction markets

Summary of technology strategy dimensional values	Actual technology strategy	Proposed superior technology strategy for niches		
	Exemplar construction company	Minor modifications	Routine new alignment	Special projects
Competitive positioning	H	M	H	H
Sourcing of technology	M	M	H	H
Scope of technology strategy	H	L–M	M–H	H
Organisational fit	M	L–M	M–H	H

Key: H – high; M – medium; L – low

There is a different debate as to whether there are opportunities for technology to create competitive advantage: in some construction sectors it is being taken up, in others it remains dormant. In 1990 Peat, Marwick & McLintock in conjunction with the CICA surveyed 900 firms in the construction industry. Although 94 per cent of the sample owned some form of computing equipment, the rate of expenditure was relatively low. Consultants spent 1.5 per cent of fee income on IT, while contractors committed 0.25 per cent of turnover. Such information begs the question of why the level of investment is so low.

The empiricist would argue that IT has not been proven to deliver competitive advantage – had it done so then managerial behaviour would have encouraged greater investment.

Gonzales et al. (1993) tackle this issue head on when they argue that IT investments need to be measured against the twin dimensions of:

* the degree of competitive advantage obtained by using IT for a particular operation, and
* the degree of necessity of that component in the business.

This perspective desegregates IT investment so that the purpose of a particular investment is considered in relation to how it will add to a business. Different parts of the firm can have different, and appropriate, levels of IT. An example outside the construction industry is that of the US Navy launching fighter bombers off aircraft carriers: these aeroplanes cost $70 million each and bristle with the very latest technology. Their movement on and off the flight deck is controlled by a man who writes with a coloured wax crayon on a window overlooking the aircraft. One part of the organisation is using

Degree of necessity of
IT component

Low   High

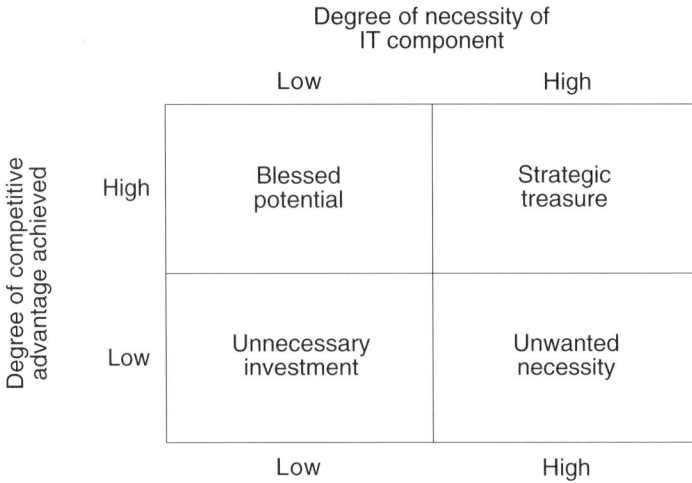

Fig. 8.5   Technology requirement in construction

technology related to the twenty-first century; the other unchanged since World War II. Both are appropriate.

Gonzales et al. (1993) created a sectored model to display different levels of opportunities for competitive advantage, as illustrated in Figure 8.5.

Unnecessary investment The box sectored low competitive advantage and low need will describe components of IT that are not needed by the organisation because they are unused or superseded by more modern facilities. For example, a design organisation with a primitive computer-aided design (CAD) system that is no longer useful for design work could be one example; a contractor with a project planning system that has been surpassed by other programmes another. Neither organisation can gain competitively by the use of such technology.

Unwanted necessity In this sector the technology may be ubiquitous in that everyone has access to the technology and so cannot create competitive advantage. Operating without this technology would handicap and so disadvantage the firm. Examples here would be word processing facilities, estimating packages, accounting spreadsheets, etc. While not creating any edge over competitors it is important that this category of IT components is managed efficiently and so costs incurred in their utilisation are minimised.

Blessed potential This sector identifies IT components that are not of themselves necessary; the firm does not gain advantage at the present. However, any investment is based on a view of the potential for the technology and that being early in the field will pay dividends. This model of early adopters of new IT opportunities requires a vision and passion for change to ensure that staff are encouraged to explore the use and potential for a new IT component. Examples here may be particular expert systems or experiments with robotics.

Strategic treasure This sector opens up the greatest opportunities for using IT to gain a competitive advantage. Not only do IT initiatives offer competitive benefits they are seen as vital to a company in conducting its business. Failure to capitalise on the IT not only will depress the effectiveness of particular operations but may render the company uncompetitive in the industry. Examples here would be major design organisations that have not adopted CAD systems in an environment where clients may be demanding the use of CAD as a prerequisite of undertaking the project. Contractors without computer-based estimating systems may forfeit tendering opportunities because of restrictions on the estimating capacity. IT designated in this zone is likely to be widely used throughout the firm and may even be customised for the specific needs of the firm.

Gonzales et al. (1993) have argued that the grid can be used by construction managers to audit their current position, which would assist policy formulation, inform investment decisions and resource allocation. Certain IT services could support a strategic move for a firm to be differentiated by cost in that IT reduces the time taken for operations or improves efficiency or by differentiating the firm from its competitors by a more extreme or unique way of employing IT. Focus strategies may be enhanced by the IT exploiting communication linkages with a narrow range of customers or services.

The Strategic Use of IT in Construction

So far in this chapter we have identified how construction managers can appraise different IT functions. Many of these functions will be routine work and it was seen that the means of using IT in a strategic way are not well developed. This is surprising given the extent and range of investment in IT. In developed countries investments in IT experienced a growth rate of 350 per cent throughout the 1980s (Keen, 1991).

The construction industry, as ever, lagged behind this welter of investment, although it does mirror the general pattern of growth. However, the application of this growth has largely been at the level of project or general administration; less is known of the strategic uses of IT in construction although examples abound in the airline, retail and banking industries of how IT has added competitive advantages. Yet these industries are significantly different from construction – all depend on linking single or multiple products to free-floating customers who can choose to buy an airline

Information technology application	Objective	Outcomes	Strategic benefits
Decision support System	To gain better and speedier decisions	More accurate estimates to the client and more efficient estimating	
Functional applications accounting, construction planning, personnel records, etc.	Rationalization of administration	Internal efficiency	
	Better management control	Higher quality	Differentiation
Construction operations	To gain information on project status	Client better informed	

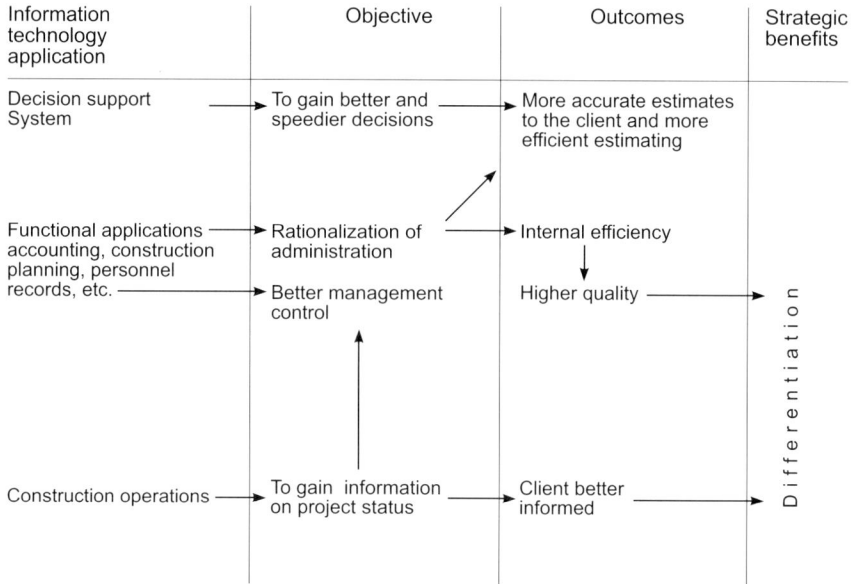

Fig. 8.6   *Incorporating strategic management into the firm*

seat or a mortgage from any number of suppliers. Construction with its project-related structure has to think of strategic applications of IT in different ways. Nonetheless, it is perhaps useful to think through an anatomy of a strategic application of IT. Ward et al. (1990) have identified four main characteristics of strategic IT systems:

- systems that link the firm with its clients or suppliers and so increase competitive potential
- systems that improve the management control of the firm
- systems that provide information-based products or services
- systems that improve productivity and performance.

Bjornsson and Lundegard (1993) examined three construction firms to see how they had incorporated IT into the strategic management of the firm. In one firm IT was used to support quality management programmes which was seen as a vehicle to differentiate the firm from its competitors. IT in support of this strategic objective was multistranded; Figure 8.6 demonstrates the linkage between IT applications and strategic benefits.

You can see from Figure 8.6 that the IT applications have sought to improve internal operations by using IT to improve decisions, automate administration and provide up-to-date project information. The strategic benefit being *differentiation*.

In another of Bjornsson and Lundegard's examples – the strategic use of IT in a management contracting and project management firm – the firm's strategic posture is to win work by cost leadership. To meet this aim, the firm has created a network of suppliers linked by an IT system. The management contractor acts as a central server for project information. This gives the firm's components suppliers access to projects even when they have narrow specialised product ranges. Designers are also within the network, so suppliers can sell to other clients through designers within the network, specifying their products.

The benefit to suppliers is that the network creates markets that enables suppliers to focus on a narrow range of products with consequent specialisation and cost reduction. The bargaining power of the supplier is obviously shaped by how easily the component may be substituted, but the managing contractor claims that by working in this way it can obtain significant cost savings for the client. This is converted into a cost leadership strategy. The impact of identikit architecture upon the aesthetics of the built environment are not reported!

Bjornsson and Lundegard's third example is a contractor whose business is based on large complex projects in an international market. Its projects demand strong technical and managerial competencies in a specialised area of business. IT is used to develop scheme designs and technical solutions in the head office which are transferred to site to be worked up into a production drawing. This process means that IT is used to communicate and distribute excellent technical resources to wherever it is required. This use enables the firm to play a focus strategy with a few international clients.

These examples serve to illustrate the linkage between IT and attempts to gain competitive advantage. It seems that IT is unlikely (in contrast to other technology) to create a permanent advantage, since the processes are easily copied and are readily purchased. The competitive advantage is likely to arise from the distinctive way the IT is combined or integrated with the business process of the firm. Consequently, the IT arenas that should be developed are the ones that contribute most to the work of the firm.

A second critical factor is the positioning of the firm in the market. If a firm wishes to work towards industry leadership it will seek to introduce an IT strategy that presents the company as one that uses IT in novel and innovative ways – but this position will have cost implications and clients may balk at the additional charges. Equally, being on the frontier can be a dangerous place and the construction industry has seen dramatic collapses of companies that have sought to be the early adopters.

Porter and Millar (1985) pose three questions to would-be technological leaders:

- Is the technological lead sustainable?
- Are there first-mover advantages?
- What disadvantages are there to being the first mover?

We'll look at these questions in turn in relation to a construction company's IT.

Table 8.6   IT applications and generic strategy

	Win work on cost	Win work by being different
Administration and support services	Cost control systems Computerised budgeting systems Office automation to keep admin staff lean	Teleconferencing with clients and designers Environmental scanning for work which is emerging Office automation to integrate admin functions
Marketing	Databases on low cost subcontractors and suppliers Economic models of regional economics building price indices, etc.	Anticipating client needs Competition analysis Virtual reality in design
Operational activity	Information applications to project control Systems performance monitoring Documenting feedback from projects Design and analysis support Inventory management Claims preparations	Information applications to research databases CAD services Quality management systems Risk assessment Bar coding Integrated construction process

Is the lead sustainable? The answer here is probably 'not for long'. As Hansen and Tatum (1989) point out, 'diffusion of technology is more widespread for basic product and process innovations than for later incremental improvements'. Leadership in the use of IT is favoured if competitors cannot easily duplicate the technology, or if the firm can repeatedly find innovative applications for the use of IT. This condition is unlikely to be satisfied in construction.

Are there first-mover advantages? Here the IT leadership can be turned into other advantages. Those with an IT advantage seldom rely on this technology alone but will convert the IT application into a more tangible benefit for clients. For example, design practices using early CADD systems defined the rules of the subsequent competition for design work; contractors pioneering the use of integrated cost and programme monitoring services to clients will have shaped expectations of services to be provided. However, to maximise first-move benefits the firm must have the necessary resources to be constantly ahead.

Are there first-mover disadvantages? Yes – often high costs are experienced. Not only is the cost of capital equipment to be borne but the learning costs of individuals and organisations can be costly. Pioneers may suffer high costs of hardware and followers may benefit from reduced costs of hardware and software. It is noticeable that many companies rely on late entry to a particular market. Proctor & Gamble are notorious for late entry only to dominate markets pioneered by others. The same is true of IBM. Coca-Cola's entry strategy is informed by the view that they allow others to develop new brand lines. They monitor the progress of the new product and then seek to muscle in by taking over the development of the brand.

By addressing these questions construction firms will be better placed to position themselves in the market(s) in which they wish to compete and have an appropriate IT strategy.

In summary, Table 8.6 catalogues the range of IT applications that can support a generic strategy of construction firms.

Market Networks and IT

These examples above illustrate how IT can create competitive advantage by developing market networks. Construction projects are classic market networks since the construction process involves the integration of the work of many individuals and firms interacting as a market network. Consequently, as Brochner (1993) points out, the construction process is the arena where the business activities of the many contributors meet and the critical task of IT is to enhance the intraorganisational and cross organisational use of IT.

The use of IT in construction usually began by automating or rationalising manual procedures rather than seeking to satisfy clients. The larger, more technically and managerially sophisticated firm will have achieved high levels of vertical integration with operating divisions in, say, property development, design, construction, facilities management, etc. It is likely that these firms will have IT applications at each stage; other firms specialising in one of the steps – for example, just the construction stage – will have limited sets of applications, and specialist trade contractors will have even less.

Consequently, when these market networks are brought together for a construction project there are likely to be gaps in the IT being brought to bear on a project. Leading firms such as management contractors or construction managers will need to bridge between firms using different levels of IT sophistication. This very participation in market networks may mean that the opportunity for the advanced use of IT is suppressed by having to travel at the same speed as the slowest in the network. High level use of IT across organisational boundaries is likely to be difficult to achieve given the independence of firms to make their own IT policies.

The introduction of co-ordinated project information in the European construction industries and work by Hutton et al. (1993) in structuring European construction data is likely to facilitate the transmission of structured intercomputer information flows.

The concept of partnering will contribute to this trans-organisational use of IT. The competitive advantages of IT are likely to be transitory since access to generic equipment, software and expertise may be mimicked by others making later investment in IT. Such investment can increase efficiency, but more importantly provide access to networks which expand business opportunities. The critical question is, how does a firm's IT strategy link with expanding the market network? In Porter and Millar's (1985) terms, it would be a cost leadership strategy with firms seeking to be sufficiently flexible to tackle work drawn from wide ranges of project and client types. This business strategy – common in a recession – will involve the firm in a large number of relatively impermanent networks with cross-organisational IT use being low. This implies generic IT solutions which can be loosely fitted to a flexible firm. On the other hand, access to networks which require a high level of cross-organisational IT capacity will require bespoke solutions with extensive investment in IT and people.

As a result the number of potential network partners is reduced and in the case of turnkey contractors (those firms that provide the client with a complete service, including land acquisition, finance, design construction, training of building users, etc.) is reduced to one firm. In short, the level of IT expected acts as a barrier to entry for certain types of projects and clients. In this setting the strategic response of the firm, in Porter and Millar's (1985) terms, is to 'focus' and the IT function is an important driver in respect of this strategy. In this setting partnering arrangements are more likely to be established and social rather than economic relations are more likely to glue the network together.

Strategically, this problem of network reduction introduces a problem; costs of IT need to be spread over a large volume of work in order to justify the investment and IT intensive networks are likely to be found where the client has a continual and large-scale building programme.

Other Strategic Applications

The above application is just one illustration of IT in the formulation of strategy; other uses can be detected. IT can be used to improve the internal efficiency of the organisation – the application here is to the internal strategy. IT can also be used for business portfolio purposes.

Internal Strategy

The typical application here is the use of Management Information Systems (MIS). Firms with effective MIS can translate this into competitive advantage.

The traditional territory for MIS has been the improvement of functional areas of the business — areas such as estimating, purchasing, accounting, project management, etc. have all found early applications.

Less well developed is the use of IT in respect of organisational structure. As construction organisations become more specialised in particular skills it is necessary to piece together 'virtual organisations made up of complex networks of the specialists necessary to design and build sophisticated buildings. In this context the business environment generates considerable inter-organisations overlaying the usual intra-organisational information. Individual managers often have major problems in coping with this complexity — humans have limits to their computational and communication abilities. IT can extend the managerial capabilities of managers; unsupported by IT the capabilities of managers are limited. The purpose of organisations is to 'marshal sufficient information processing and communication capabilities, to be able to manage the complexity and uncertainty inherent in the environment' (Bakos and Treacy, 1985). IT will enable more rapid communication (e-mail, faxes on site, etc.), which closes the physical and organisational gaps between the design and construction functions in settings where these are conducted by separate organisations. In more modern integrated structures the internal efficiencies will be assisted by the communication between functional specialists such as designers and cost estimators.

The second area where IT can serve the internal structure may be observed in the way information is processed. Here the MIS enhances construction managers' capabilities to receive information that is filtered to identify critical aspects of, say, project performance. For example, a site manager will need to know performance of activities on the critical path in advance of subcritical works. He or she will need to know the exceptional items which are running over budget, etc. These informational nuggets are the output of information processing in order to enhance the performance of managers in the two dimensions of communication and processing of information. IT, through MIS, will do three things to enhance the work of managers. It will:

- extend the *capacity* to communicate and process information
- enhance the *quality* of the communication (often by using coloured graphics) and the information processing
- reduce the *unit-cost* of communicating (for example, by using disks rather than paper) and processing (for example, by cannibalising designs through CAD systems).

Business Portfolio

The organisation structure of firms in the industry will also be informed by Williamson's (1975) markets and hierarchies theories. Most construction

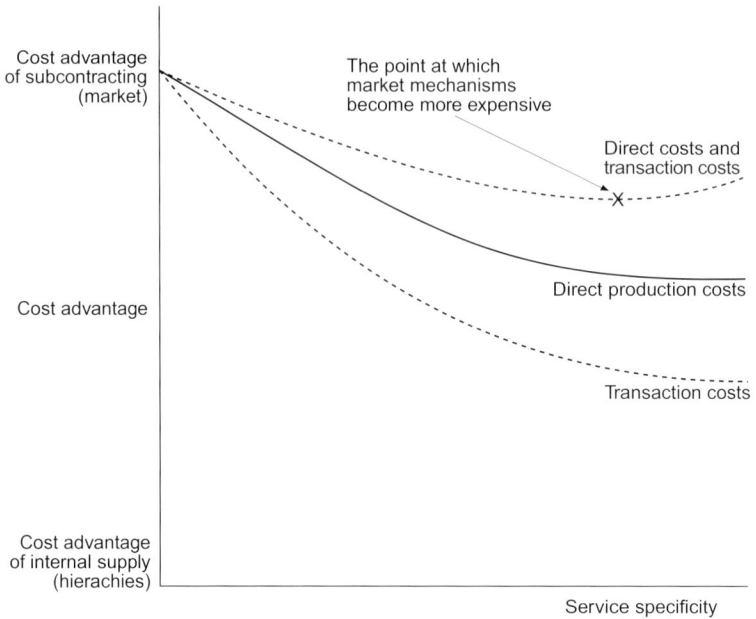

Fig. 8.7   Comparisons of costs of direct labour and sub-contracting

contractors will have choices to make between operating in the market for production resources (typically subcontractors) or establishing hierarchies by providing the service inside the organisation. The conventional wisdom has been that cost competitiveness is best served by subcontracting since the transaction costs of providing construction resources in-house is seen to be prohibitive. Williamson (1975) sees the cost advantage of operating the market mechanisms declining as the specificity of the product or service increases. The relationships between internal provision and subcontracting is shown in Figure 8.7.

The critical factor is to determine where x falls. In construction the level of specificity is likely to be low for everything but the more specialised and technical of trades or project management service. However, the application of IT can affect service specificity by reducing the price of transaction costs by creating internal efficiencies through, say, electronic data processing.

Equally, a firm could affect the production cost advantages of using a subcontractor market by transferring information technologically rather than manually. In both cases the economics of undertaking work directly or subcontracting have been changed by the intervention of IT.

Future IT Application in Strategic Management

The IT revolution has prompted many changes in terms of the construction process and the product (the completed building). However, the impact of this revolution has been patchy. Some industries, such as financial services and airlines, have been turned upside down by the impact of IT. Construction, as a whole, is less affected. The influence of IT will be variable between firms and between functions. For example, the quantity surveying function may be IT rich since the very business of quantity surveying is concerned with generating and processing information. Other design professions may have different levels of evolution from sophisticated firms with fly-through computer graphics to entice clients, to those who operate with drawing-board and drafting pen. The same disparity is experienced by construction firms. Large powerful contractors will have experience of electric drawings distribution and 3D modelling systems, while trade contractors in basic building crafts will use more primitive applications.

What one may say with confidence is that the environment for construction which contains the work of all of these firms will continue to be turbulent and so forecasting and predicting will continue to be difficult. This suggests that environmental scanning will become important, with an ability to sense weak signals coming from the environment being a critical skill.

Consequently, construction firms will need to be agile corporations: to quickly respond to rapid changes in the construction environment.

To be agile will require a constant flow of information from internal and external sources to allow informed decision making to take place. IT provides the information systems that will improve the environmental scanning and the co-ordination of a response.

The internal information will need to be integrated to obtain the benefits of modem management techniques such as total quality management (TQM), materials planning and delivery with just-in-time (JIT) methods. These integrated IT systems will improve internal efficiency and will promote closer links between different functions within the firm.

The external information will have to be evaluated for its accuracy and its impact on the business of the firm. Research undertaken within Strathclyde University has sought to use IT to model this information to create a construction industry futures bulletin. The next chapter describes this innovation.

References

ALSHAWI, M. and AOUAD, G. (1995). 'A strategic integration of information technology and business strategies: a structure methodology'. *Journal of Civil Engineering Systems*. Vol. 12, pp. 249–261.

ANSOFF, I. (1985). *Corporate Strategy*. New York: Free Press.

BAKOS, J. and TREACY, M. (1985). 'Information technology and corporate strategy: a research perspective'. CISR WP No 124, MIT.

BETTS, M., LUM, C., MATHUR, K. and OFORI, G. (1991). 'Strategies for the construction sector in the information technology era'. *Journal of Construction Management and Economics*, Vol. 9.

BJORNSSON, H. and LUNDEGARD, R. (1993). *Strategic use of IT and Construction in Management of Information Technology in Construction*. Singapore: National University of Singapore.

BRANDON, P., BASDEN, A., HAMILTON, I. and STOCKLY, J. (1989). *Expert Systems: The Strategic Planning of Construction Projects*. London: QS Division of the RICS.

BROCHNER, J. (1993). *Construction Process Improvements in Construction Networks in Management of Information Technology in Construction*. Singapore: National University of Singapore.

CONSTRUCTION INDUSTRY COMPUTING ASSOCIATES and PEAT, MARWICK & MCLINTOCK (1990). *Building on IT for the 1990s: A Survey of the Information Technology Trends and Needs in the Construction Industry*. London: Peat, Marwick & McLintock.

EARL, M. (1989). *Management Strategies for Information Technology*. London: Prentice Hall.

FLAATEN, P., MCCUBBREY, D., O'RIORDEN, P. and BURGESS, K. (1989). *Foundations of Business Systems*. Florida, CA: Arthur Anderson Dryden Press.

GALLACHER, J. (1988). *Knowledge Systems for Business*. London: Prentice Hall.

GONZALES, A., OGUNLANA, S. and SOEGAARD, R. (1993). *Technology Impact Grid: A Model for Strategic IT Planning for Competitive Advantage in Construction in Management of Information Technology in Construction*. Singapore: National University of Singapore.

HAMPSON, K. (1996). 'Technology management in construction: a management, teaching and research framework'. In Langford, D. and Retik, A. (Eds), *The Organisation and Management of Construction*, Vol. 3. London: E & FN Spon.

HANSEN, K. and TATUM, C. (1989). 'Technology and strategic management in construction'. *Journal of Management in Engineering*, Vol. 5, No. 1, January.

HUTTON, G. H., MCGREGOR, D. R. and MACLEOD, I. A. (1993). 'An intelligent interface for construction information'. Paper presented at ASCE 5th International Conf. on 'Computing in Civil Engineering and Building', California.

JAMES MARTIN & ASSOCIATES (1988). *Information Strategy Planning*. London: James Martin & Associates.

KEEN, P. (1991). *Shaping the Future*. Cambridge, MA: Harvard Business Press.

MRAOVIC, B. (1996). 'IT as a function of managerial strategy in the organisation and management of construction'. In Langford, D. and Retik, A. (Eds), *The Organisation and Management of Construction*, Vol. 3. London: E & FN Spon.

PORTER, M. and MILLAR, M. (1985). *Competitive Strategy*. New York: The Free Press.

RIBA (1992). *A Strategic View of the Profession*. London: RIBA Publications.

WARD, J., GRIFFITH, P. and WHITMORE, P. (1990). *Strategic Planning for Information Systems*. New York: John Wiley.

WILLIAMSON, O. (1975). *Markets and Hierarchies, Analysis of Antitrust Implications: A Study in the Economics of Internal Organization*. New York: Free Press.

Chapter 9

An Application of IT to Strategy Formulation in Construction Firms*

Introduction

This chapter looks at the application of IT to the exploration of the multiple faceted business environment that may be faced by construction firms.

Often IT has been seen as a tool to assist project management, in terms of time control, monitoring of construction progress, etc. Frequently it is used to support design through computer-aided design (CAD) and more recently visualisation of the product of construction and the process of how it may be built. Early applications for using IT was as support for administration, such as record keeping, wages and salary calculations. However, it has not been until recently that IT has been able to be applied to an environment that combines quantitative data that needs qualitative judgements to be made about the data.

This chapter records an experiment in which choices and preferences are modelled by the use of sophisticated IT. The setting of these choices was one whereby strategic decision makers in construction need to explore the alternative business environments for the construction industry. These convey multiple possibilities and by modelling them the strategic planner may lay plans for the firm with greater confidence. It discusses a product produced by a research project undertaken at Strathclyde University that produced a piece of software call The Construction Alternative Futures Explorer (CAFE).

The Question of Turbulence

One of the characteristics of the construction industry is that its business environment is highly turbulent. The volume of work available to consultants and contractors fluctuates widely. To illustrate this, Figure 9.1 shows the

*This chapter was prepared in conjunction with J. Brightman (Banxia Software), C. Eden and K. van der Heijden (University of Strathclyde).

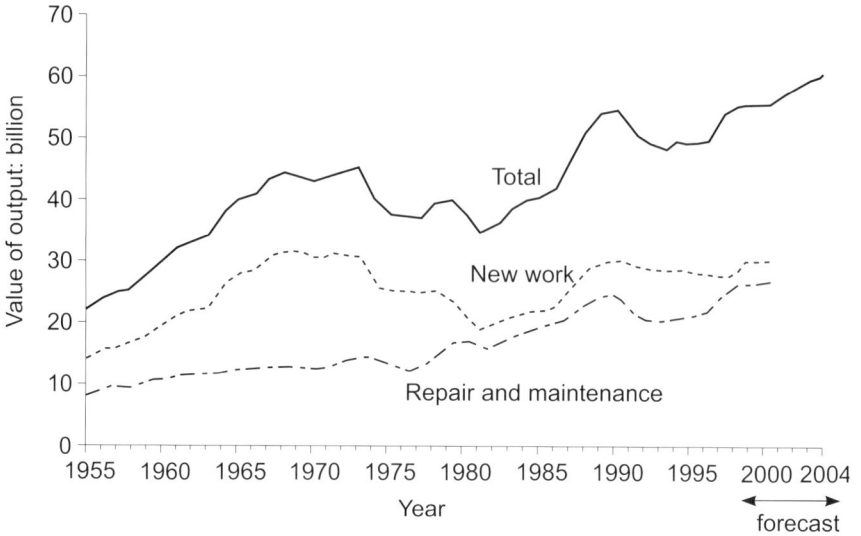

Fig. 9.1   Value of construction output 1955–1999

marked changes in the value of orders for new construction work over the period 1955–1998 at 1990 prices.

This variability of demand has led to fluctuations in employment opportunities in the industry, as illustrated by Figure 9.2.

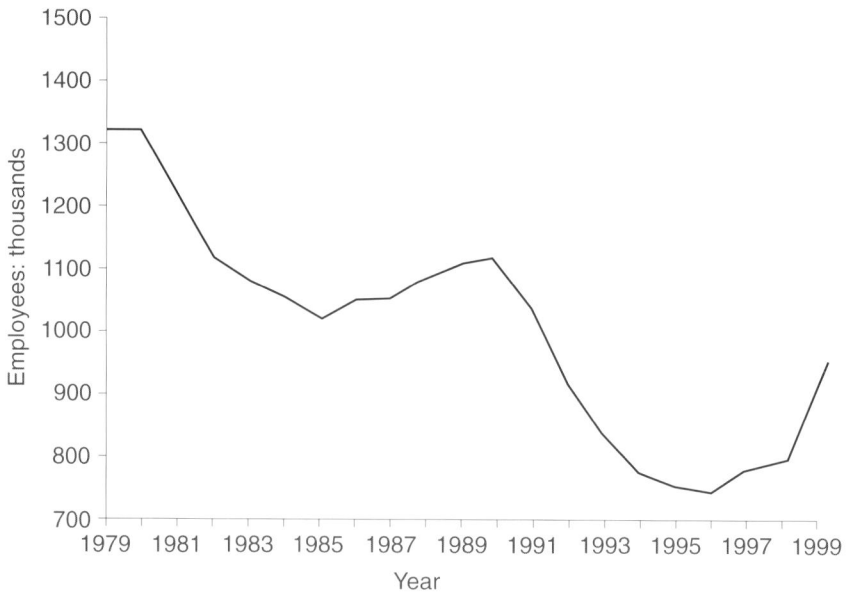

Fig. 9.2   Number employed in construction 1979–1999

Another feature of the industry is that its structure is largely based around regional companies, with different regions experiencing different variations in their levels of construction activity. It is a fragmented and hierarchical industry (Male and Stocks, 1991). Using the south-east of England, the West Midlands and Scotland as examples, it can be seen that the new orders obtained by contractors was strongly skewed in favour of the south-east of England in the period 1986 to 1998, with steep falls in demand after 1989. In comparison Scotland does not share the boom but equally does not slip into as deep a recession. This is illustrated in Figure 9.3. The low points in the business cycle stimulate more fierce competition among contractors for all types of construction work.

The need to operate in this environment has consequences for the modus operandi of construction firms. One response has been to create business plans in an ad hoc rather than a formal, strategically planned manner. This ad hoc approach enables firms to respond to rapid shifts in the business environment. Miles and Snow (1978), in creating a typology of organisational strategic responses, see firms falling into one of four categories: the defender, the prospector, the analyser or the reactor.

The defender seeks to sustain competitive advantage by being cheaper than competitors. This response is well fitted to stable environments with well-established and fairly predictable patterns of demand. The prospector thrives in more chaotic settings, where change in the environment is rapid and unpredictable. These very changes throw up opportunities, either in new markets or in new services delivered to established markets. The analyser

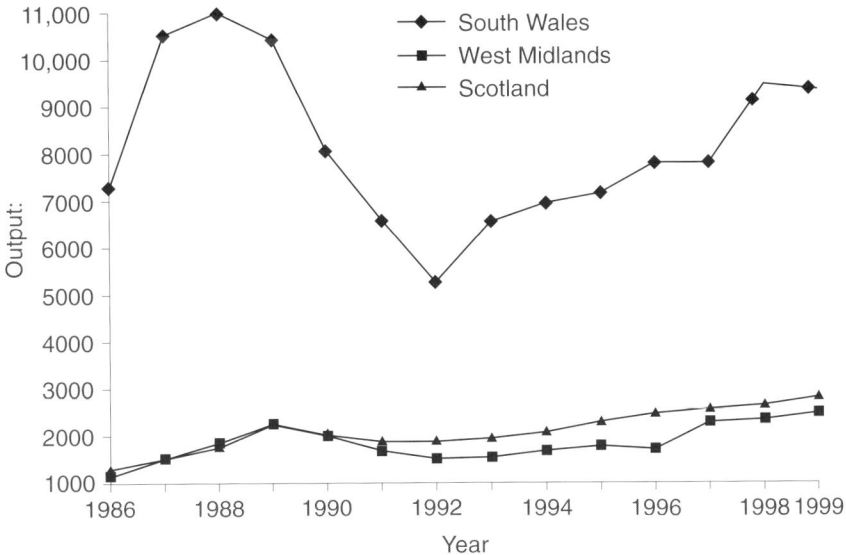

Fig. 9.3   *Output in 3 regions at 1990 prices*

blends the caution of the defender and the adventure of the prospector. The analyser seeks to research changes in the environment and may set up special arms of the business to behave in ways typical to the prospector. Finally, the reactor, who is seen to spin and turn in the wind, reacts to changes in an inconsistent and unstructured way. The belief that construction organisations were characterised by this final type of behaviour set the scene for a research project carried out at Strathclyde University. The research sought to investigate the nature of strategy formulation in construction firms and subsequently explore ways of assisting the industry to manage alternative futures in a more strategic manner

Planning in a Turbulent Environment

Strategic planning implies an ordered systematic process which creates a direction for a company to follow. It involves analysing the company in terms of its strengths and weaknesses and analysing the opportunities and threats which exist in its operating environment.

In theory this analysis is then converted into policies, plans, tactics and other activities necessary to achieve the plan – this process implies an ordered outer world and a relatively static hierarchy within the internal structure of firms. This model is sometimes appropriate for business environments which are dominated by a few large firms but it is unhelpful in settings where there are many firms and all are able to offer similar services and may frequently only be differentiated by tender price. In volatile business environments traditional strategic planning technologies are seldom useful. In the circumstances of the construction industry strategic planning has to be elevated to strategic management. This is a step beyond strategic planning. It is:

> a set of managing practices that link the day-to-day operations of an enterprise to the planning and decision making that must accompany longer time horizons. Below the surface, in its fullest meaning, the term strategic management suggests a state of affairs in which all members of a dynamic organisation move as one in response to plans made, opportunities and threats. (Ansoff, 1987)

The dynamism of the environment demands strategic management techniques which fit the nature of the firms in the industry and the characteristics and organisational culture of its managers.

Within this setting one approach that is often helpful in promoting practical strategic thinking is scenario planning. This approach enables firms to consider multiple futures, which could, in turn, be used to inform and improve strategic thinking. Scenarios have been used as a means of informing strategic thinking in a number of different settings such as the energy and automotive industries (Jenkin, 1993, 1995; Sviden, 1991; Wack, 1985a). More recently the UK government reported a Technology Foresight Programme, which considered potential future scenarios for the whole of the construction sector, identifying those science and technology developments which might have a significant impact on the construction environment. This was done with a view to stimulating future research activities in the appropriate areas.

Scenarios are designed to recognise that firms find it difficult to forecast using trend extrapolation – the forecasts are generally wrong because unexpected events throw the trend off course. Thus it becomes important to think about the future from the stance of a number of possible futures, each of which may be the particular combination of important events capable of significantly influencing the industry.

Langford and Male (2001) define scenarios as having three characteristics. They are:

- *Hypothetical* – they expose us to several different possibilities for the future.
- *Vague* – they do not give detailed descriptions but provide a generalisation of what may be.
- *Multi-disciplined* – they provide an overview that brings together many aspects of a society: technological, economic, political, social, demographic and environmental.

The scope of a scenario will depend on what is attempted – a global scenario will be wider in the impacts it considers than a national scenario. Equally an industry scenario will be wider than one for an individual firm. The appropriate level is determined by the strategic setting and mission of the particular organisation.

There are a number of different scenario methodologies in use, for example:

- *intuitive logic* (Schwartz, 1991; van der Heijden, 1995; Wack, 1985b) – this depends on views of the future being based upon a 'feel' for the future by senior managers or 'experts' who 'scan the horizon' to sense what is coming up
- *trend impact analysis* – this explores historic trends and extrapolates them to the future with an identification of break points or 'paradigm shifts' in the direction of a trend
- *cross impact analysis* – this is a technique for combining several trends or events. A strategic shift (say a change of political control in government) will have a cross impact upon a very wide range of social and business activities. A smaller but more obvious one – say a change in the bank base rate – will promote or choke off investment plans. Cross impact analysis gauges the impact of one event on others

Futurology methods include the consideration of powerful actors (Eden and Fischer, 1993; Eden and van der Heijden, 1996).

The approach adopted for the research to be described in this chapter is closest to the intuitive logic method. This type of scenario relies less on probability and more on qualitative causal thinking. As such it seems to appeal more to the intuitive needs of the typical decision makers (van der Heijden, 1995). 'Event based' scenarios – narratives made up of interconnecting events – were used, and in order to build such scenarios a carefully constructed set of rules, about which events should be collated, were generated.

The Construction Industry Model

The research programme moved through three phases. The first phase was aimed at establishing current practices within the industry and exploring the kind of information that construction managers draw on to inform strategic thinking. The second and third phases, which ran in parallel, addressed the use of information technology in supporting strategic thinking and the use of scenarios in a construction context. The overall structure of the work is depicted in Figure 9.4.

The first phase revealed the information environment in which construction managers work. They use a variety of information sources from industry statistics to industry magazines focusing on opinion from experts and pundits. Information is drawn from inside and outside the firm. What was notable was the ways in which organisations source information for purposes of strategic thinking. The wealth of government data, most frequently presented in statistical form, was seldom used by firms; the reason given being that such data needs time and effort to convert it into 'digestible' information. Only one firm in the sample used government statistics and even this firm was sceptical about their usefulness. Consequently, firms tended to rely upon secondary information – that which had been converted by forecasting organisations or firms providing gazettes for future projects. The professional and trade press were seen as particularly helpful in delivering 'state of trade' type reports which would lead to an appreciation of the robustness or fragility of the overall market.

Discussions held with contractors in this phase of the work confirmed that construction firms are 'strategically mobile' and opportunistic; to quote one managing director, 'if it's bridges next year, we'll be in bridges'. This strategic flexibility leads firms to describe themselves as being demand led, relying on clients to initiate projects rather than generating demand for their projects; in short, 'we can't plan, so we don't plan'. The planning which was evidenced was detailed short-term (3–6 months) planning, with workload and turnover targets being articulated for a 2 to 3-year planning horizon.

Our sample showed little evidence of formalised and highly structured strategic planning, a view which is supported by other studies (Wayment,

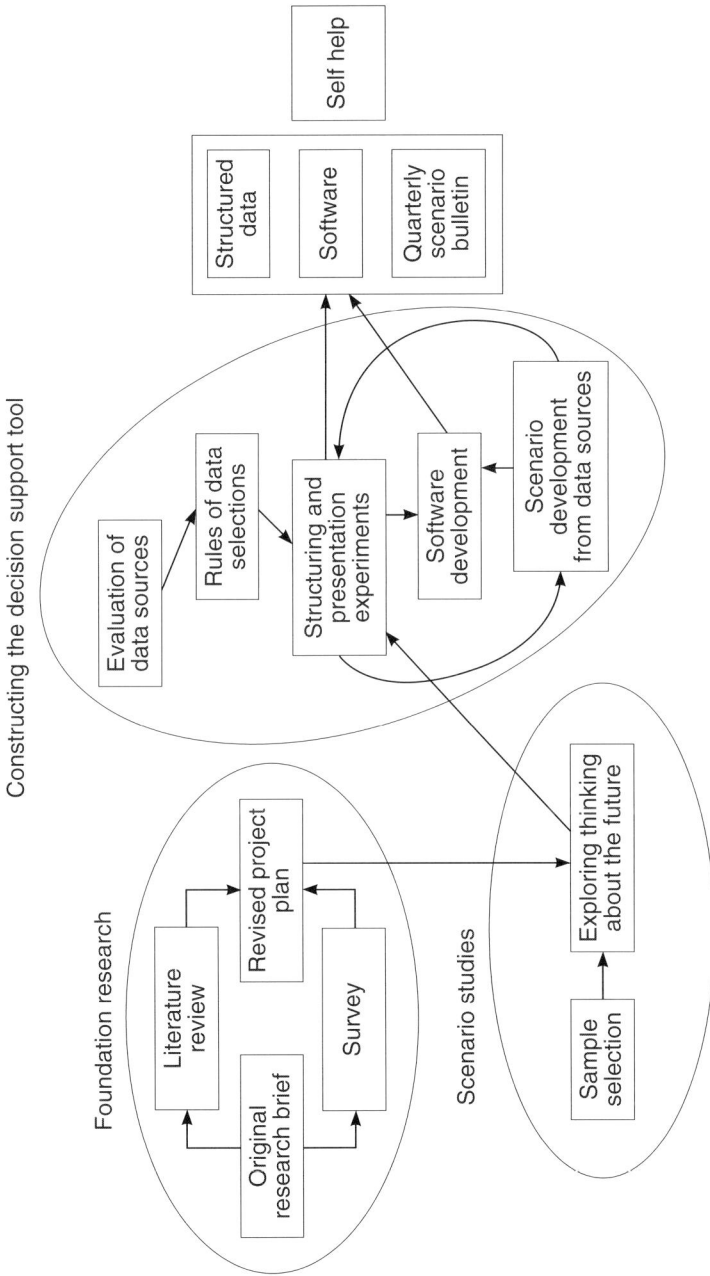

Fig. 9.4   Structure of data collection

1993; Hillebrandt, 1987; Hillebrandt and Cannon, 1990). Hillebrandt and Cannon also note that the management of projects rather than the firm is the centre of managerial attention. Like most managers, our sample relied heavily on soft information, including personal contacts with architects and other design professionals as useful sources of strategic intelligence.

During the second phase of the research, scenario studies were undertaken to establish the differences between what can be provided through general data extracted from pubic data sources and the individualistic demands of companies. In the third phase information technology was used to assist with the presentation, manipulation and structuring of strategically significant data in order to develop scenarios. On completion of these parallel activities the final stage of the research was to test the system with the managers of a construction firm.

The first modelling experiment, to construct alternative futures based on events predicted by industry pundits, involved extracting 'headline' type information from 18 trade publications. These publications were representative of some of the sources, providing strategic intelligence, which had been identified by the managers interviewed. Some of the information extracted from these publications was general, widely known intelligence, while other information was speculative in nature. The information was presented in map form, a map being a series of nodes and linking arrows. The nodes of the map (referred to as concepts) are short phrases representing events in the business environment. These concepts were distillations of critical factors in a future scenario and are drawn from information sources which the earlier survey had marked out as being used to inform thinking about the future. A map, which is a section of the model, is shown in Figure 9.5.

Initially, the concepts were classified by the research team into one of three groups:

- whether the event spoke of construction market activity and so had business relevance to a firm because it affected the future activity base (activity statements). For example, a government announcement of a capital spending programme.
- whether the event added to the firm's understanding of the capabilities it would have to acquire to make an impact on the market (capability statements). For example, if a firm had to acquire BS5750 to tender for public-sector projects.
- supporting commentary, historical or contemporary, putting events in context.

These divisions were used to enable the effective use of colour graphics, so that attention could be drawn quickly to the most significant events. The activity and capability statements appear at the top of the model and are supported by events which are 'commentary' and 'history' — this material traces the sequence of events leading up to the activity and capability

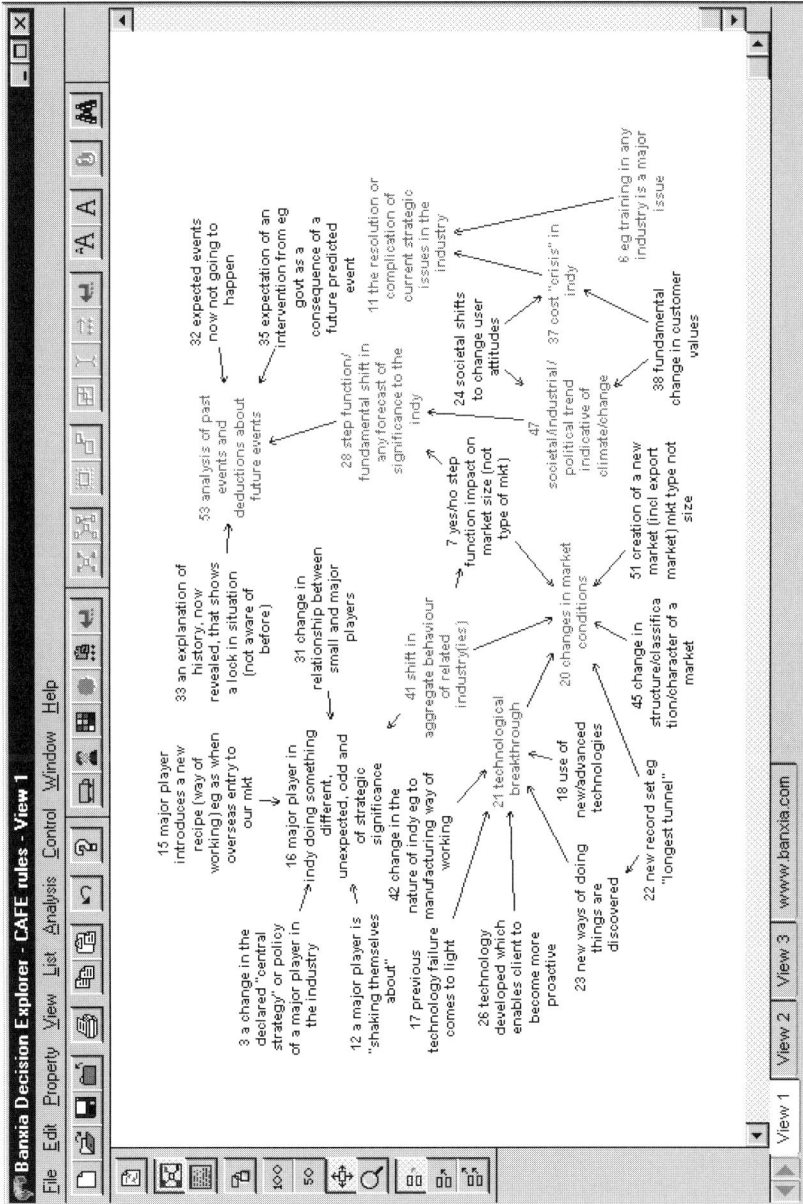

Fig. 9.5   CAFÉ scenario map view (a starting point for map exploration) (courtesy of Banxia Software)

statements. Behind all of the concepts were memo cards providing the source reporting the event and in many cases amplifying the 'headline' information.

These concepts were then modelled using the Decision Explorer software. This process involved making linkages between the concepts. Some of these linkages would be causal. Simply put, one story could have related to interest rate increases and another to further falls in the housing market – clearly these could be linked together. In the end we had created a model of the construction industry.

The central concepts in the model were those which were more frequently linked to other concepts. If a concept, say interest rate changes, triggered off many other activities to shape the capability of firms or the activity in the market, it was a key concept in the model and so likely to shape the business behaviour of firms in the industry.

For example, construction market forecasts from the JFC Forecasts would be a useful source of information and 'headline' stories from forecasts would be initiated. Equally, a news story which is likely to influence strategic thinking in the industry would also be abstracted. For example news stories of major players withdrawing from the housing market will be a story of scenario significance.

The culmination of later experiments was the model containing approximately 500 events, with a complex network of links. Consequently, it was helpful to use a series of analysis routines which could determine the emergent themes in the model. Through a combination of detecting key themes, central events that fall out of the analysis as being at the core of many possible futures, the computer software was used to create scenarios for the future which construction organisations could use in planning their business. The end product is a decision-making aid which we called an industry futures bulletin (IFB). The IT used organises information, both soft and hard data, to create multiple pictures of the business environment in which firms may have to live and exploit.

A small sample of firms agreed to become more integrally involved in the research and an IFB was built in the expectation that it might inform their strategic thinking. This model was tested with one of the companies. The purposes of these experiments were:

(i)   to see whether this approach to introducing data about strategic futures really had any impact on the thinking of practising managers

(ii) to explore the way in which it would be used, so that appropriate software interfaces could be designed.

The experiment started with the managers of the companies looking at the busiest themes in the model, those events which had the most connections to others. The executives then 'walked through' the model by exploring the themes which were of greatest interest to them. By exploring the model the managers chose one of several pathways and during this process the exploration of the model stimulated discussion of the firm's strategic thinking. The software served to focus thinking on strategic issues – it had some impact on mentally tuning the thinking of the executives.

One of the expectations in carrying out the scenario studies was that there would be a significant overlap between the data gathered from the previously identified publications, and used as the basis for the computer model, and the issues raised in the scenario workshops. An interesting outcome was the relationship between the 'big map' of the industry as developed by the research team and the narrower range of issues explored by the executives. Some individuals participating in the scenario workshops appeared to exhibit what could be called 'blind spots'. For example, they had difficulty in conceiving of new forms of competition and it was also extremely difficult to elicit views of a 'bad future' especially when a certain issue had been identified as being important to the firm and yet was unpredictable in its outcome.

Inevitably the companies explored the model in the way which was most relevant to their needs and fitted their own perceptions of their capabilities and scope. Our intention is that users should be able, and encouraged, to insert company-specific data in order to place their own specific circumstances in the strategic context provided by the model.

The positive response of the collaborating firms suggests that the product of this research does have potential industrial use. Whilst the construction industry is not generally noted for its participation in management research or its use of ''traditional' strategic-management principles this research has sought to bridge the gap between management theory and construction-management practice. The system appears to be a usable and flexible means of influencing managers' thinking about future business environments and trying to encourage thinking about alternative futures.

The research has produced evidence that it is possible to enhance futures thinking amongst managers in the construction industry (Brightman et al., 1997). By locating scenarios on activity and capability issues the method focuses on the building up of distinctive competencies and capabilities in anticipation of possible, and possibly new, demands from the market. The IFB delivers decision support in assisting with this process, but so far this requires facilitation through strategy-development workshops. However, the research provides an example of how IT can be successfully used to marshal and order data in support of strategic management in construction firms.

Further work is ongoing (Brightman et al., 1998) to refine the method of entering and exploring the model and to add a help system so that independent users are guided through the model and prompted to add their own data. The use of memo cards also needs to be more clearly defined so that the additional information which is given in them is 'what the managers want'. The researchers are encouraged by their results to date and are continuing this work with the ultimate aim of developing a stand-alone decision support system which could be used by organisations without expert facilitation.

References

ANSOFF, I. (1987). *Corporate Strategy*, 2nd edn. London: Penguin.

BRIGHTMAN, J., EDEN, C. VAN DER HEIJDEN, K. and LANGFORD, D. (1997). 'Construction Alternative Futures Explorer (CAFE) – User Guide. Strathclyde: University of Strathclyde.

BRIGHTMAN, J., EDEN, C. VAN DER HEIJDEN, K. and LANGFORD, D. (1998). 'The development of CAFE'. *Journal of Automation in Construction*, Vol. 8, pp. 613–623.

EDEN, C. and FISCHER, G. (1993). 'The role of modelling cognition for stakeholder analysis and scenario planning'. International workshop on managerial and organisational cognition, Brussels, Belgium, 13 and 14 May.

HILLEBRANDT, P. M. (1987). 'The management of large UK contracting firms – theory and practice'. In Lansley, P. R. and Harlow, P. A. (Eds), *Managing Construction Worldwide: Volume 2 – Productivity and Human Factors in Construction*. Proceedings of the 5th International Symposium on the Organisation and Management of Construction. London: CIB-W65, 7–10 September 1987, pp. 919–26.

HILLEBRANDT, P. M. and CANNON, J. (1990). *The Modern Construction Firm*. London: Macmillan Press Ltd.

JENKIN, F. P. (1993). *Energy for Tomorrow's World*. New York: World Energy Council and St Martins Press.

JENKIN, F. P. (1995). 'An overview of world future energy demand'. *Nuclear Energy, the Journal of the British Nuclear Engineering Society*, Vol. 34, No. 1, February, pp. 21–6.

LANGFORD, D. A. and MALE, S. P. (2001). *Strategic Management in Construction*, 2nd edn. Oxford: Blackwell Science.

MALE, S. (1991). 'Strategic management in construction: conceptual foundations'. In Male, S. P. and Stocks, R. K. (Eds), *Competitive Advantage in Construction*. Oxford: BSP.

MALE, S. P. and STOCKS, R. K. (Eds) (1991). *Competitive Advantage in Construction*. Oxford: BSP.

MILES, R. E. and SNOW, C. C. (1978). *Organisational Strategy, Structure and Process*. Tokyo: McGraw Hill.

REPORT OF THE TECHNOLOGY FORESIGHT CONSTRUCTION PANEL (1995), Cabinet Office, HMSO. London.

SCHWARTZ, P. (1991). *The Art of the Long View*. New York: Doubleday.

SVIDEN, O. (1991). 'A scenario method for forecasting'. *Futures*, Vol. 18, No. 5, pp. 681–91.

VAN DER HEIJDEN, K. (1995). 'Scenarios: thinking about the future'. In Rosell, S. A. et al., *Changing Maps*. Ottawa: Carleton University Press.

VAN DER HEIJDEN, K. (1996). *Scenarios: The Art of Strategic Conversation*. New York: John Wiley.

WACK, P. (1985a). 'Scenarios: uncharted waters ahead'. *Harvard Business Review*, Vol. 63, No. 5, pp. 73–89.

WACK, P. (1985b), 'Scenarios: shooting the rapids'. *Harvard Business Review*, Vol. 63, No. 6, pp. 139–50.

WAYMENT, M. (1993). 'How can UK civil engineering contractors become more strategically orientated?' Unpublished and confidential MBA dissertation, Brunel University, Uxbridge.

Chapter 10

Conclusions

What's next?

Internet and Integration are probably two the most important agents of change in the first decade of the Millennium. There are many examples of how the synergy of combining these trends has already generated novel approaches and efficient solutions which were unthinkable just a few years ago. It is interesting to watch also how having realized the huge potential and benefit of Internet and Integration, many competitors are forming partnerships and working together to advance the implementation of such solutions.

Two most notable examples of such initiatives that are described below are interrelated. In our view, these examples demonstrate how computer and communications technologies are affecting the Construction Industry in a similar way the Industrial Revolution had changed the textile industry and introduced the whole concept of manufacturing and mass production. The first initiative – Web Based Enterprise Management – is an Internet based infrastructure for allowing creation of applications for managing distributed enterprises. The second initiative – Industry Foundation Classes – presents an approach that takes advantage of and allows to capitalize on such environment.

WBEM (Web-Based Enterprise Management) – an industry initiative that establishes management infrastructure standards and provides a way to combine information from various hardware and software management systems. WBEM specifies standards for a unifying computer system architecture that allows access to data from a variety of underlying hardware technologies and software platforms on Internet, and presents that data in a consistent fashion. Management applications can then use this information to create solutions that reduce the maintenance and life cycle costs of managing an enterprise network. The WBEM standard includes the Common Information Model (CIM), which is an industry standard driven by the Distributed Management Task Force (DMTF). DMTF has about 200 member companies including leading hardware and software companies such as Cisco, IBM, Intel, Microsoft, Sun and many others (see more on *http://www.dmtf.org*)

The International Alliance for Interoperability (IAI), is a public not-for-profit organization, which was formed initially in 1994 by a group of leading American AEC companies as an Industry alliance. Within a few years it came to be an International movement counting over 600 members world wide. This is an industry-led initiative as opposed to STEP (which is an ISO initiative), however, it gradually became to be de facto standards, uniting a wide range of the organizations from architecture and engineering to construction and facility management. The organization's mission is to define, publish and promote specifications for Industry Foundation Classes (IFC) as a basis for project information sharing in the building industry (architecture, engineering, construction, and facilities-management). The information could be shared across all disciplines and technical applications throughout the project life cycle.

To achieve this, IFC defines a single way of describing buildings in computer systems. Based on an object oriented data model, such description allows data exchange and information sharing by all IFC-compliant applications. IFC is an open and public model, and it evolves in time as the need arises. It can be used by any member of the Alliance for its own implementation (see *http://iaiweb.lbl.gov/* for more information).

The communications technology rapidly advances in utilizing Internet and wireless networks for interconnecting sites and companies using relatively inexpensive personal computers, which are now virtually available on every site. This trend creates enormous opportunity for the construction industry to capture and efficiently exchange and share the vast amount of information generated for each building project that is usually lost. Avoiding duplication and increasing interoperability of software applications are the key factors for a more efficient and modern construction industry. This is a win–win situation for everyone.

Appendix A

Cases of IT Implementation: Evolution or Revolution

Introduction

The following case studies are derived from real cases where computerisation was applied to organisations. These successful examples of IT implementation demonstrate a range of views, approaches and solutions. Bearing in mind different backgrounds, goals, timing and different technological solutions available for each case at a time, the descriptions have been focused on the implementation process and decision making rather than on the technological solution.

    CASE I: Planning and Management: a client approach
    CASE II: Design: a consultant approach
    CASE III: Construction: a contractor approach
    CASE IV: Integrated Design and Construction: a concurrent engineering
                approach
    Conclusions for all cases

Case I: Planning and Management – a client approach

Adapted from Dr Peter Bingham (Director of Information Systems Strategy Unit, Cheshire County Council), 'A view on council information system strategy in a local authority'. Unpublished seminar paper, Institution of Civil Engineers (ICE), London, 1994.

Background

Cheshire County Council is a large organisation with a gross budget around £1 billion per annum. It delivers a wide range of services to the local community including education, social services, highways and transportation, libraries, police, fire and so on. Up until the end of the 1980s the services provided by each department appeared to be separate and distinctive: the organisation was departmental in its approach. The result for IT was development of separate and unrelated computer systems. Initiatives to develop corporate or council-wide applications were fraught with difficulty and susceptible to failure. The consequence on the user was a degree of duplication of effort.

The driving computerisation force at that time (1980s) was the perceived need to automate administrative and clerical processes. These were mainly finance based: payrolls, payment to creditors, superannuation, etc. The beneficiaries of this expansion were users and managers located in financial departments.

Other departments which could not get access to this mainly corporate IT provision 'split off' from the county main frame. Departmental systems were installed in highways, police, architects, social services and libraries. Individual members of staff invested in micro computing.

Also, county-wide IT was seen as being driven by the finance people. It was unrelated to any IT planning in departments, with consequences on the planning process. The IT department based its business on what is now seen as traditional data-processing projects: large-scale developments which required highly specialist staff to develop and implement. The service was not charged directly to the clients but was limited by the resources allocated to the IT supplier.

Cheshire's Approach

Following a managerial review in 1989, it was realised that the problem of excessive departmentalism would need to be addressed, as would some of the underlying causes: strategy for IT was technology driven; budgets were held at the centre by the IT provider and bids for projects were assessed by a central panel. There was little evidence of an information-systems strategy for each of the major services. The council also needed a council-wide strategy which would focus more on *what information is required* to support the business of the council rather than an IT strategy which has a focus on *how information* services are delivered.

No single methodology was identified as being suitable to deal with this practical problem. Cheshire could be described as a federal organisation; The separate groups had a great number of delegated powers to develop service strategies and to manage delivery. The corporate role (both in business terms and in terms of IT) was therefore to make sure that the council had a medium-term view of where it was heading, that it set standards and to make sure that the service departments owned their strategies and plans, based on their own business needs.

This was reflected in the management structure with the formation of an information systems (IS) planning function, separate from the delivery of IT services, with responsibility of an internal contractor. The planning outfit represented the clients. The contractor worked in an internal market with products and services fully charged. All market budgets were distributed to clients.

The various reviews of information delivery confirmed the need to solve the problem of systems which were separate and not co-ordinated in any way. Following the reviews, priorities were set within a mission to 'facilitate the transfer of information to any person, anywhere, in support of service objectives, recognising confidentiality and security'.

In relation to the mission statement, the common feature which became a core part of the IS strategy was development of improved personal support for office-based staff. Common demands emerged which focused not on applications but on the need for individuals to communicate messages and documents quickly and easily; and for managers to have easy access to a common set of well-supported information products and services. The networked personal computer was identified as the key device, with menu-driven facilities which gave (a) access to electronic office services, (b) departmental information systems, and (c) information services.

This changed the emphasis in IT delivery. Whereas in the past the mainframe computer and departmental machines were seen as the focus for IT, the personal computer on the desk became the key hardware feature which would deliver the strategy to the user. The first step in achieving the vision was to replace what had been duplicate communications networks with a combined voice and data communication facility.

A product, CHESHIRElink, was planned to provide, at the desk, a range of facilities which included:

- Personal computing
- Office systems
- Statistics, members' address files, maps, facts and figures
- Financial information
- Personnel information
- Property information
- Access to departmental systems.

This approach, won the British Computer Association/Hay Management Consultants national award for excellence in IT in 1992. It can be summarised as follows:

- IS strategy and IT strategy are always considered as separate but related issues
- neither the model of a central monolithic IT structure nor local autonomous units are appropriate
- a synthesis between the two is required and a balance must be struck between them based on specific circumstances and organisation.

Conclusions and Lessons Learned

A key feature of Cheshire County Council's approach to information systems is that the responsibility for each group's IS strategy and plan lies within the group. The group strategy informs the County IS strategy and is informed by it. The County IS strategy itself is not a blueprint: it is an annual process of renewal.

Clients own their strategies and development plans. They hold their own budgets and purchase services under internal-market arrangements which provide a degree of security to both the client and internal contractor.

Top management at board (council) level does not exercise any control over tactical plans. Its focus is on high-level strategic issues.

Finally, the strategy delivers a product. CHESHIRElink is a manifestation of a strategy which sees the county council as a networked and inter-communicating organisation. It does so by means of standard components of hardware, user interfaces, communications, e-mail and word processing which can be delivered to any location in the county. These components can be added to or re-configured as computer applications for the benefit of all subscribing users.

CASE II: Design – a consultant approach

Source: Adapted from P. Rutter (Partner, director of Scott Wilson Kirkpatrick's IT Systems Group), 'A partner's view on the continuing evolution of IT'. ICE Seminar, London, 1994.

Background

The Scott Wilson Kirkpatrick (SWK) group is a large international consultancy. It provides services for civil and structural engineering, transportation and environmental planning. It has offices in 24 countries. The Scott Wilson Kirkpatrick Group's IT systems support the work of 750 staff in the UK and 2000 world-wide.

Computerisation History

SWK's early commitment to the computer dates from before 1970. Initially operating structural analysis and computer-aided drafting systems from a bureau computer, the firm soon installed a central computer system in its head office.

Following expansion of the firm in the late 1980s the central computer was devolved to the larger regional offices, to which their branch offices were linked.

By 1992, the central VAX computer system, having undergone several upgrades, had been replaced by clusters of VAX processors located on LANs in the main regional offices and connected by BT Kilostream links to the Basingstoke head office. A growing number of PCs using bought-in engineering and project-management programs were operating as stand-alone hardware.

The Current IT Strategy

An IT strategy formulated at the beginning of the 1990s to place a networked PC on the desk of every engineer within five years was thwarted by the recession. In the event, the more gradual introduction of PCs has resulted in an investment in hardware which is more powerful and versatile than the VAX processors and in network software which is less prone to problems than earlier products on the market. With hindsight the maxim applicable to the world of IT, 'never buy today what you can put off buying until tomorrow', has proved to be correct.

AutoCAD software is now widely used throughout the firm for computer-aided draughting. It is increasingly linked to other software, for example the CAD analysis, design and detailing software for structures. The most up-to-date software available for design and analytical tasks is mainly PC based. Networking allows software to be shared between a number of PC stations located within an area accommodating a particular discipline at a lower license cost that would apply if a license for a centralised computer system had to be purchased.

ISDN lines are now being used to provide digital connections to smaller offices, and to replace Kilostream lines where usage is intermittent. Links are being established with the offices world-wide with a view to achieving standardisation and compatibility of systems and procedures so that efficiency is improved and so that staff moving between offices will be familiar with the facilities. Electronic mail links with overseas offices are being introduced to reduce the cost of fax transmissions.

IT Planning and Budgeting

The annual budget for computing hardware and software leasing and maintenance is currently 1.7% of the UK firm's annual turnover. A further 1.3% is attributable to the central IT department staff costs, consumables and communications links.

For the partner responsible for IT in a multidisciplinary engineering firm there is a continuing balance to be maintained between responding to the requests for new hardware and software from all sectors of the organisation, both technical and administrative, and the financial constraints of the annual budget allocated to IT. The difficult factor in this financial equation is quantifying the cost benefits in improved efficiency which the purchase or leasing of new hardware and software achieves. While the drawdown on the available budget is all too apparent, the improved productivity may not be so discernible.

Previously, computer equipment was leased over five years and the useful life expectancy was reckoned to be at least of comparable duration. Present-day life expectancy of hardware is not determined by its durability but by its suitability to support the latest software. Consequently, the leasing period chosen is now three years which is more in line with the expected frequency of renewal of equipment.

The increasing demand for PCs and portable PCs for job-related tasks has led to the establishment of a pool of hardware which can be 'rented' out to the project and returned to the pool after use. This is intended to prevent the accumulation in a particular location of equipment which is under-utilised when the project task is completed.

Conclusions and Lessons Learned

The most significant change over the years, apart from the power of computers, has been the move away from in-house program writing to the buying-in of software. There remains the capability to write application programs but this activity is undertaken largely by the IT department staff. In earlier years, there was a substantial commitment to the development of engineering programs but the commercial viability of doing so is doubtful in view of the large number of specialist software houses marketing engineering and management programs.

Whereas a few years ago the demand for computing tended to originate from specialist staff, the request for availability of hardware and software now comes from staff at all levels. This stems from two reasons; first, the larger selection of software for project-management tasks and design functions, and, second, the growing user-friendliness of the software. A further factor is the number of staff whose education has been undertaken in an IT-aware environment and who have progressed to middle and senior management.

Standardisation should be sought so that part of the system can communicate with another and one office with another. This is particularly important with text file transfer.

The principal lesson learnt from the past investment in computing has been that evolution rather than revolution should dictate the pace of change. The justification for expanding the IT system into new areas must be clearly made and the cost benefit assessed before incurring the outlay. There should be resistance to purchasing new hardware and software unless such acquisitions fit into the planned development of the organisation's IT system. IT is essential for the successful operation of a business. It can save money and waste it. Only by careful management of IT, in accordance with a strategic development plan which has the support of the partners or directors of the firm, can the benefits it provides contribute to profitability.

CASE III: Construction – a contractor approach

Adapted from D. R. Jones (Group IT Director, John Laing), 'Achieving a major IT change in a large civil engineering contracting company', ICE Seminar, 1994, London; and 'Introducing and managing IT in a civil engineering business', Proceedings of Institution of Civil Engineering, Paper 10588, 108, 1995, pp. 84–7.

Background

John Laing is a large, UK-based civil engineering contractor taking part in many complex and prestigious projects. Yet, until 1991 most of the company's computer activities were centralised and mainframe based. Not only was the computing service expensive but it was not meeting the company's business needs. The decision was taken to create a new IT infrastructure appropriate for the 1990s, and implement it as soon as possible. Laing required a completely new IT strategy that was driven by the future business plan, had a vision of five years, delivered major business benefits, and gave far more empowerment to the company.

Implementing the new infrastructure was a major five-year IT project which delivered a flexible infrastructure of about 70 computer networks with more than 2,000 computers connected together in one large wide area network. The overall cost of IT has been cut by one third (over £4 million savings!) and business managers rather than the centralised IT department are now responsible for the running and operating of most of the applications.

Laing's Approach: Revolution not Evolution

Introducing a new IT infrastructure and moving from a centralised IT bureaucracy to decentralised divisional computing is a painful process. In order to succeed, seven basic requirements were identified.

Agreed and shared vision

The first requirement for any major IT project must be an agreed and shared vision so that everyone from the top to the bottom of the organisation understands and agrees with what is being done and why. A shared vision needs senior management commitment and involvement.

Agreed long-term IT strategy

The second requirement is to ensure that the vision leads to an agreed long-term IT strategy and is driven by the long-term business strategy. The strategy must contain the costs and benefits of the IT project as well as schedules for delivery. The benefits must be demonstrable and traceable. This strategy must be 'sold' first to the senior management and then to the rest of the organisation in a series of events (that were called 'roadshows' at Laing).

Involvement of the whole organisation

The third requirement is to involve the whole organisation in the complete cycle of IT implementation, with a definite split between the technology requirements and the business/end-user requirements. It is imperative that the IT part of the organisation is only involved in the technology decisions and kept well apart from the business decisions. This technology/business split is difficult to handle correctly in that the two are heavily entwined if an overall 'big picture' solution is to be found.

The John Laing approach was to set up a broad Common Technical Architecture (CTA) and agree that no software systems were to be developed in-house and all chosen software had to comply with the broad CTA. John Laing then had end-user groups evaluate software applications and choose three short-listed vendors for each major category. The company evaluated software covering six major categories and ended up with 18 short-listed vendors. Only one of these vendors had to be rejected because of non-compliance with the broad CTA.

Once the final choice in each category was made, the software application choices and the standards contained within them on relational databases, graphical user interfaces and others led to the redefinition of a detailed CTA with which the technical project(s) could proceed.

Business process re-engineering

While evaluating the various software applications, the opportunity arises to review and to question how the company operates and what can be done to improve existing processes and procedures.

Business partnership with IT suppliers

The fifth requirement is to seek partnerships with the chosen IT suppliers in the hardware/software/network products so that a complete 'team' culture can be created for the actual implementation. Partnership between all the various companies involved in a major IT project is critical if the project is to be successful.

Decentralisation

The sixth requirement is to decentralise some aspects of computing. In this a radical approach is seen as desirable.

If possible the existing set-up should be frozen and the 'new' should effectively be a green field site. At the same time agreement will be needed on the conformity of the technical infrastructure. The new infrastructure should be as simple as possible. In the area of decentralisation, care should be taken to ensure that in moving away from a centralised approach certain central systems that benefit the organisation overall are not lost. It is essential that a common approach to systems management, security and disaster planning are agreed.

Changing from centralisation requires diplomacy and tact to ensure the 'centre' acts as a co-ordinator in gradually handing over responsibility to the working divisions.

Delivery in time and within budget

The last requirement is to deliver what was promised to schedule, in time and within budget. To do this, the project needs to be managed professionally while utilising technical and commercial skills to ensure that the best hardware/software choices were made.

The project managers will also have to ensure that, during the IT planning and implementation phases, use is made of all the project-management skills and tools normally associated with large-scale projects. It is the failure to use these controls that is the major reason for large IT projects failing.

The Laing Experience

The John Laing 'signed-up vision' was a three-phase IT project covering five years, 1993–1997. The *first phase* was to freeze the current IT environment and replace it with a new improved IT environment over a period of two years. The *second phase* was a one-year period of stability to allow the new IT environment to be enhanced (tuned). The *final phase* was to re-visit the strategy and add those business-value systems that were required for the future.

After the initial scepticism of senior management and lack of end-user commitment and understanding of the requirements, the project received and continues to receive full commitment from the main board. The response from the divisions in taking on their new responsibilities was described as 'outstanding' by the staff responsible for the project and the business partners and internal IT technical staff were said to have provided 'excellent support'.

After the implementation of the first two phases the IT profile within Laing is described by those involved as 'very positive'. The project has kept within budget and the implementation has kept ahead of schedule. Decentralisation has occurred without any loss of sensible central requirements, such as central payments or central suppliers. The project has established a new solid IT infrastructure for the whole organisation that is suitable for the rest of the decade. The new business systems have demonstrated greater functionality and additional business benefits. Overall annual IT costs have been reduced by £4 million per annum. The foundations have been laid for even greater benefits for future IT projects.

Conclusions and Lessons Learned

Most of the problems associated with either introducing new types of information technology or making major changes to an organisation's IT relate to the fact that the IT fraternity can forget who is the client. One of the major reasons for this is that the client refuses to take ownership of IT. The major objective of IT must be to have the IT strategy driven by the business needs; a technical infrastructure must be implemented to support systems that are driven by the future business strategy.

The other objective of IT must be to add value to the business process. This value must be demonstrable, traceable and effectively add to the business's bottom-line profit. All of this can only be achieved by creating a partnership of trust between the staff and IT that actually works.

CASE IV: Integrated Design and Construction – a Concurrent Engineering Approach

Source: Adapted from J. V. McManus, 'Implementation of concurrent engineering in the offshore industry'. International Conference on Concurrent Engineering, London, Institution of Structural Engineers, 1997, pp. 1–9.

Background

Kvaerner Oil & Gas Ltd's experiences of imposing pressure to reduce costs and shorten the duration of offshore development was not good. Such pressure resulted in parallel working; increased interdependence between subcontractors often led to conflicts resulting in contractual claims and subsequent escalation of project costs.

The company realised that an improvement was required in contractual relationships to eliminate inefficiencies at the project interface. The clear challenge was to develop integrated project teams based on a more co-ordinated approach. A major objective was to create seamless interfaces between all contractors. Concurrent engineering techniques were a major component of this objective.

The Concurrent Case

The company's main area of operation is the production of oil platforms for the North Sea oilfield. A typical offshore platform consists of topsides facilities for production, utilities, and living and drilling services which are supported by a piled steel jacket. Major projects recently constructed have installed weights of 20,000 tonnes for the topsides facility and 7,000 tonnes for the substructure.

The offshore project schedule (see an example in Figure A1) is about three and a half years long. All activities for design, procurement, construction, commissioning, installation and offshore hook-up are completed during this period, after which the project is handed over for production services.

Detailed review of the schedule demonstrates the concurrent nature of the task. There is a need to commence both procurement and fabrication activities very soon after the start of the design period. At the point of fabrication award (i.e. the procurement stage), the design will be only approximately 45% complete.

The fabrication and commissioning schedules require the early commitment for purchase of a large amount of the primary steelwork. The delivery of steel must be made early in the design programme and before detailed definition of equipment is available to the designer.

ACTIVITY	Year 1				Year 2				Year 3				Year 4			
	Q1	Q2	Q3	Q4	Q1	Q2	Q3	Q4	Q1	Q2	Q3	Q4	Q1	Q2	Q3	Q4
Preliminary Engineering	▨	▨														
Detailed Engineering			▨	▨	▨	▨	▨	▨	▨							
Procurement			▨	▨	▨	▨	▨	▨	▨							
Fabrication & Transport					▨	▨	▨	▨	▨	▨	▨					
Nearshore Hook-up & Tow												▨				
Offshore HU & Commission													▨			
Drilling													▨			

Fig. A1   A typical schedule for an offshore project

Typically, the structure contributes 51% of the weight content, while other important components are mechanical (16.5%), electrical equipment (9.5%) and piping (13.6%).

The concurrent nature of the schedule for design and procurement has in the past been a major source of inefficiency . The failure to correctly size the low-percentage equipment commodity by adding design margins will result in cost penalties in the structural steel take offs, with price impacting not only on materials but also on the fabrication processes and installation. It is therefore essential that the flow of information from the supplier is aligned to the design process at an early stage.

The fabrication programme is dependent upon the delivery of all material and equipment in advance of construction needs. Failure to deliver will severely affect the schedule if the procurement and fabrication interfaces are not aligned correctly.

The following critical areas are identified as major areas for consideration:

- alignment of the design and fabrication schedules
- the scheduling of information requirements from suppliers in support of the design programme
- a procurement schedule that ensures delivery of all materials in advance of construction needs
- an understanding of the fabrication methods by the design team, thus avoiding re-work at later stages
- the scheduling of testing and commissioning requirements that is acceptable to the fabrication programme
- the input of the operating team in the design to ensure that all life-cycle requirements are considered at the initial stage.

Implementation

Despite an initial reluctance to break away from the protection of the discipline group, a multidiscipline method of working was created. The teams focused upon key objectives and developed a wide understanding of the end goal. The culture of the teams became supportive, reducing conflict and blame. Problems were reviewed collectively and a team approach was adopted in seeking solutions. The multidisciplined and multiskilled approach in the project environment has provided a major improvement at the discipline interface.

Communications and integrated systems

The company used documents and drawings as the main communications media in the design office. It was found that much of the information was repeated in many areas. This was a source of delay and increased operating costs. It was found that different groups within a project have similar data needs, but each produced or acquired data in different formats. These differing formats, especially those handwritten, caused ambiguity and misinterpretation. By using a central database and linking the information, it was found that the information flow was vastly improved.

The establishment of interactive systems turned serial activities into parallel systems, eliminating repetition. For example, information could be passed automatically between disciplines; engineering data, such as material take-offs, were immediately available for purchase orders. Similar links established allowed planners to directly receive accurate and current data immediately from all areas of the project, providing improved control systems.

This reduced paper, review periods and information handover times within the project team.

Extending the database availability

The next step was to extend access to the database to other contractors. By electronically linking the company design team with operators and fabricators interface activities have been greatly improved. By sending all data via a network link, documentation is immediately available at the construction workface. Providing access to the design database has permitted the fabricator to use the same drawings as the designer, converting the design into a status suitable for fabrication. A recent example gave savings of four hours on each piping isometric drawing. This saving was made on a total of over 6,000 drawings producing a considerable reduction in time. This example was repeated in many areas, enabling control of concurrent activities and actually achieving cost reduction.

Data management for life

The principles are now being expanded into the operating life of developments. Access to the central database in the design office will be maintained during the operating phase of projects via satellite communication links. This will provide instant access to all information that has been generated in the project life and eliminate repetition of activity.

Conclusions and Lessons Learned

Successful control of the concurrent engineering process has required a number of changes to working practices. Many lessons have been learned and, in particular, the following actions have contributed to successful project execution.

- acceptance of the need for cultural change, in particular the use of multidisciplined work groups and multiskilling

- alignment of project schedules of all sub-contractors to ensure that the respective objectives of all project team members are supported
- the sharing of design databases to eliminate repetitive work and save time
- the maximum utilisation of IT systems with particular use of electronic transfer of documentation between locations
- improved contractual relationships to eliminate traditional adversarial interfaces.

However, the most important lesson learned was to recognise and understand the problems that concurrent engineering created and to seek solutions. This led to a major change in attitude in the way to approach projects and, as a result cost, and schedule reductions have been achieved. The elimination of many of the inefficiencies has resulted in improved performance in many areas.

The recent project has provided the company with a benchmark for the future: it was completed half a year quicker with total savings of 18%.

Conclusions for All Cases

These studies present real cases of comprehensive implementation of IT in various types of companies with different types of activities. All demonstrated very different approaches to strategies of implementation. Nevertheless, there are very important and fundamental similarities. They are:

1   The long-term IT strategy should be derived from the needs of the business and its vision of the future. It will be unique for each organisation and should, if appropriate, integrate any organisational or cultural changes within an organisation.
2   The strategy should endorse a bottom-up, top-down approach; it has to be driven by those who use IT; it must address their needs and be truly supported by management.
3   IT implementation should be carefully planned, phased and adhered to.
4   Education and technical training should be provided not only to users but also to managerial staff. A mix of seminars, workshops and presentations will help to change attitudes, generate enthusiasm and teach new skills.

Appendix B

Cases of IT Implementation Using the Internet

Introduction

There has been little experimentation by the construction industry in the use of the Internet (or intranets and project-specific intranets in particular) for company and project management. The following three case studies describe how intranets have been used by some companies and some of the benefits gained.

Case V: Civil engineering experience
Case VI: Contractors' experience
Case VII: Project management experience
Summary

Case Study V – TransCanada Pipelines Ltd

TransCanada Pipelines Ltd is a natural gas and crude oil transmission, processing and marketing company. It is based in Calgary, Alberta and has a turnover of around $US12 billion. A typical project for the company is moving oil or gas along a pipeline across North America, and its problem was the co-ordination of projects where teams were geographically dispersed.

Director of strategic direction at the company, John Mackenzie, said: 'Projects are easy. Teams are difficult. It's a question of how fast we can get 12 to 20 key people working like a well-oiled machine' (Wilkinson, 1998). These key people are spread across TransCanada's 50 offices in Canada and several in the United States, Mexico and South America. The company also has very tight deadlines, set by the government, and unless the deadlines are met government approval will not be forthcoming.

The company felt it needed something more practical than voice-mail or e-mail, so its IT department standardised departmental intranets and linked them all together.

The result was a corporate intranet, from which they could run all projects, using an application called Livelink. This application allows teams to be built more quickly, management is more efficient, projects are completed in less time, and less money is spent on corporate travel, phone bills and paper consumption. Livelink also allows projects to be planned in tandem, collaborating bulletin boards, task lists, project library and workflow and audit trail.

The intranet was also used on a project where TransCanada linked with two other companies to build a pipeline between Chicago and Dawn, Ontario. This project was called the Vector Pipeline and was valued at $40 million. The major problem with the project was the tight deadline of five months, set by the Federal Energy Regulatory Commission.

The project incorporated parties in Houston, Minnesota, Toronto, Connecticut, Chicago, eastern Ontario and Calgary. TransCanada also had two of its Canadian offices working on the project. Mackenzie's fear was that that the project would have problems because setting up and co-ordination of teams usually took six to eight months, but using Livelink the team was able to be co-ordinated without setting up a head office and the government deadline was met.

Savings on the project were evident. The paperwork of the Federal Energy Regulatory Commission would have been hundreds of pages long; all of this was written, updated and changed many times by using Livelink. It was also transferred to federal headquarters in Washington electronically. It was estimated that, by doing this, the paperwork took half the time anticipated and the cost was slashed from $12 million to $6 million.

TransCanada is currently in the middle of a merger with another Alberta-based pipeline company. If this merger is successful the users of Livelink will grow from 3,800 to 7,800 users.

Source: Wilkinson, 1998.

Case Study VI – ICA Fluor Daniel

This company specialises in the engineering, procurement and construction of industrial projects. Its aim in using an intranet was to 'enable global work teams to co-ordinate, collaborate and communicate more effectively on projects, thereby increasing productivity and profitability' (http://home.netscape.com/comprod/at_work/customer_profiles/ica_fluor.html).

The solution was an intranet that 'would enable global sharing of project information'. The intranet facilitates access to company policies, regulatory databases and complete ISO9001 procedures from anywhere in the world. The ability to access ISO9001 saved $25,000 on paper costs in only six months.

The addition of Prospero an intranet-based application which allows people to share work while being in different places, gives ICA Fluor Daniel a project forum for world-wide collaboration and communication which will reduce complexity and costs.

The business benefits of using the intranet, for ICA Fluor Daniel include:

- improved productivity of global work teams
- reduced costs
- better customer service.

The company plans to extend the existing system to incorporate an extranet to allow clients, suppliers, sub-contractors and other key parties to have access to project information, allowing greater collaboration between all of the parties.

ICA Fluor Daniel estimates that it saves hundreds of dollars a day by the reduction in courier services and the elimination of delays through lack of information; these savings being achieved by the use of the intranet.

Source: Netscape, www.netscape.com/comprod/at_work/customer_profiles/ ica fluor.html

Case study VII – Fentress Bradburn Architects Ltd

This Denver-based architectural firm implemented an extranet – an internet that allows access by people outwith the project team so long as they have security clearance. The purpose of this extranet was to assist in the design and management of the construction of a business centre. In particular, the aim of the extranet was to assist the project team with the project management and collaboration and communication with the clients.

As with the case of the corporate Intranet case studies, the firm has now embraced the extranet as its new way of carrying out its project management and communicating. Using the extranet the client needs only a web browser on his or her PC to review the progress and gain a better understanding of the project. Without an extranet, a client would have little involvement in the project and not know what to expect from it on completion.

The extranet allows outsiders access to a 'shared space' where two people can access the same information at the same time. This would be very useful when the client is using the extranet; the client would not affect the work whilst looking at the project information; the contractor or designer can look at the information at the same time.

The company sees the extranet as its way ahead; it plans a complete switch to using the extranet for all work and communication with clients, and plans to use the extranet for all projects in the future.

A company spokesperson has said: 'We believe this is a better way to manage a project. We see technology as one of our key advantages, and we use it to improve our practice and the architecture practice at large' (Intranet Design, 1998).

For this extranet service the company only has to pay a small fee at the outset, well within the project's budget. The rest of the fee can be paid on a monthly 'subscription' fee or the company can buy the system outright. The system was set up in one visit to the architect's office by the extranet team, which caused minimal upset to the project operations or resources.

The architect picked this system as the company had worked with the same information technology company in the past; its quality of work had been more than satisfactory.

The extranet system unifies all communications, schedules and design files, generated with one system to be compatible with most other file types. The system will also associate all drawing files with related text files.

This system was designed specifically for the use of Fentress Bradburn Architects Ltd, and they have embraced its use, in the same way that Ford has embraced its intranet. The company has offices in Denver and Oakland, and the intranet will help reduce a lot of travelling and postage between the two offices and any construction sites the company has. The extranet is used for projects ranging from $3 to $500 million and the project types vary from public-sector buildings and office buildings to airports and museums.

Through using an extranet Fentress Bradburn Architects Ltd. have gained some obvious benefits, such as increased client access to project information which will lead to increased client satisfaction. Up-to-date information will lead to fewer mistakes and rework, thus reducing the overall project price and contract length. The case study, while not containing any specific financial information, does suggest that the initial investment was quite small and the intranet was quite affordable.

Source: Intranet Design, www.innergy.com/bin/newscashe?item = 395

Summary for Cases V–VII

Intranets are not being used widely in construction-related disciplines. This may be because of a refusal by professionals in these disciplines to be proactive instead of reactive. Many people will need to see a problem before any action is taken, however proactive professionals would see the potential benefits of an intranet without such a problem forcing them to make a decision on the issue.

The examples of Internet-based management of construction companies and projects are positive. The cost of set up and running is not significant, but the benefits are many.

CAD Glossary

3D Three-dimensional – a display, medium or performance giving the appearance of height, width and depth.

agent A software entity (usually a part of graphical user interface) that can perform a task (simple but tedious) for the user, e.g. search, retrieval, help.

algorithm A procedure for solving a particular problem.

alpha version The initial release new (software) product to be tested before selling. *See also* beta version.

animation A graphic method where the illusion of motion is created by rapid viewing of individual frames in a sequence. Fifteen frames (images) per second is a minimum to produce a continuous movement illusion. Many software packages (e.g. 3D Studio) will provide facilities to produce an animation from 3D models.

application program A computer program developed to support 'real' human tasks, i.e. calculation, writing, drafting, as opposed to tasks of operating of the computer itself.

artificial intelligence (AI) Emulation of the cognitive aspects of human behaviour, e.g. decision making and learning using computers. AI is one of the computer science disciplines. It includes expert systems, knowledge-based systems, decision, support systems, pattern-recognition systems, cybernetics and others.

artificial reality *See* virtual reality.

ASCII American Standard Code for Information Interchange. This name is given to the computer character set established by the ANSI (American National Standards Institute).

augmented reality A virutal reality technique of viewing computer-generated images super-imposed on real-world views.

backup A safety and/or achieving measure of the duplication of data files on an external measuring device (e.g. disk, tape, CD).

bandwidth The capacity of a communication channel showing the volume of information that can be transmitted within a given time (e.g. TV images require a bandwidth of 6 MHz).

beta version The release of new (software) product to customers who agree to use the product prior to general distribution and to report problems (if any) in their applications. This release follows the alpha version.

bi-ocular Displaying the same image to each eye; done to conserve computing resources, when acceptable.

bit Binary digit – a small unit of data that can be stored in a computer.

bitmap A computer screen image comprised of dots (pixels) that correspond directly to data bits stored in memory.

browser An interface to a database that allows the user to quickly scan the database content.

buffer (a) In scheduling, an interval between two linear activities to avoid collisions; (b) in computer systems, a memory allocation reserved for temporary storage of data.

bug A defect in computer software.

bus A channel along which electronic signals pass from one part of a computer or network system to another.

byte A number of bits grouped as a unit for processing and storage. Although six, seven and nine bit bytes have been used, the eight bits byte is now used virtually by all computer systems.

C Programming language widely used for software development because of its ability to provide concise and efficient programming code.

C++ Object-oriented version of the C language.

CAD Computer-aided design – the application of computers and particularly computer graphics to the engineering design.

CAE Computer-aided engineering.

CAL Computer-assisted learning – the use of computer technology for educational and training purposes.

CAM Computer-aided manufacture; when manufacturing process and operation of machinery is operated by computers; sometimes provided by CAD output.

CD-ROM Compact disc read only memory – it is an adaptation of audio compact disc technology to computing; a disc holds up to 640 Mb of digital data.

compression The reduction of the amount of data required to store and transmit information.

data compression The term for techniques to reduce the size of a computer file for storage or transmission purposes.

data Known facts that can be recorded and that have implicit meaning. It is common to use the word *data* in both singular and plural.

database Structural collection of data organised in such a form that manipulation and retrieval are possible.

DOS Disk operating system.

EDI Electronic data interchange – a networked telecommunications service allowing organisations to exchange electronic documentation taking us closer to the 'paperless office'.

electronic publishing A term for publishing in electronic form (e.g. CD-ROMs, on-line, publishing).

e-mail Electronic mail is a system that allows messages to be sent from and received by computers via networks.

emulation Modelling behaviour of another computer system.

ethernet A local area network system developed by Xerox in the 1970s.

expert system A computer program that makes specific expert knowledge available to other users through emulating the performance of a human expert; it contains encoded statements, 'rules', that represent individual or gathered expertise and can perform reasoning applying 'inference engine' to reach decision or come to a conclusion, etc.

fibre-optic transmission Sending a large amount of data as light pulses through glass fibre; an expensive but accurate, reliable and fast medium.

floppy disk A removable electro-magnetic storage medium.

FTP File transfer protocol – a method of transferring a file from one computer to another, provided both are on the Internet.

full duplex Two-way simultaneous communication, e.g. telephone.

GIS Geographical information system – a computer map-based information system.

GPS Global positioning system – a computer-based navigation system that calculates a user location on the Earth's surface using satellite signals.

groupware Computer package designed to assist individual users to work together as a group through a computer network.

half duplex One at a time two-way communication, e.g. fax, modem.

hardware Electronic components and equipment of any computer system, including network links between the equipment at different locations.

head-mounted display (HMD) Helmet or head-held brace with optical or electronic display devices located in front of the user's eyes.

hypertext A program that allows users to create cross-references and links (hyperlinks) between different sections and even locations of information in a random manner.

Image A manual, mental or computer-generated reproduction of the appearance of real or abstract object or objects; a picture or drawing that is usually perceived as being pixel based.

immersive VR A virtual reality system creating the illusion of being immersed in a synthetic world using a head-mounted display and position tracing sensors linked to the user's body.

information Data processed in such a way as to be of some meaning to its recipients.

interactive Enables the user to manipulate or influence the course of the action according to the user's choice.

interface A piece of electronic equipment to connect between computer and external devices or a program that provides communication between the user and computer program or between the different programs and software.

Internet A super network which interlines virtually all networks creating the largest computer network in the world.

ISDN Integrated Services Digital Network – a worldwide public telecommunications network. Being digital rather than analogue (*see* PSTN, modem). ISDN is a quick and efficient way to receive and transfer large files, such as drawings, graphic images, video and voice, thus providing an infrastructure for video-conferencing and similar facilities.

knowledge Understanding how to access, interpret and use information.

LCD Liquid crystal display – a type of flat screen. An electronic field applied to a surface of liquid crystals causes them to become so oriented that they act as polarised light filters.

modem MODulator/dEModulator – an electronic device that converts digital signals from a computer into analogue signals to be transmitted through the telephone system, and vice versa.

multimedia Combined text, images, full-motion video and sound; requires lots of bandwidth and computing power.

object-oriented Conceptually grouped into autonomous units or structures.

paradigm An example or conceptual model used to illustrate new approaches or way of thinking.

parallel interface An interface permitting transmission of data in a parallel way, i.e. several bits at a time (usually eight) using one wire for one bit. Majority of printers use this type of interface.

parameters Measurement factors or determined bounds.

process Defined activity or sequence of actions or executions to reach a result; the repeated execution of the logical tasks identified in terms of input/output; *not* procedure.

program This is a coded set of instructions that interprets the information given to the computer with the keyboard, mouse or any external device and then directs the computer to carry out a task.

prototyping A technique for building a quick, rough version of a desired system or its parts to allow analyses and evaluation of the concept or to explore system possibilities.

RAM Random access memory – primary electronic memory in a computer where programs and data are stored while the computer is working; can be overwritten; provides fast access to its information but loses the information when the computer is switched off.

real time The actual time something occurs; in computing it defines type of programs that provide response to input fast enough to affect subsequent input.

reality engine Computer system with software for generating virtual objects and worlds and enabling user interaction.

Rendering Translation into another form, e.g. converting signals into picture; performing pixel-oriented calculations for display; yielding or reducing state or interpretation.

resolution Measure of picture quality, usually represented in dots per inch (dpi); more dots give sharper image.

RGB The primary colours: red, green and blue; the combinations and intensities of these colours are used to represent the whole spectrum by colour displays.

ROM Read only memory – data storage built into a computer's hardware to hold such information as the computer's operating system. These data can be retrieved by the processor but cannot be written or replaced.

serial interface An interface permitting transmission of data in a serial way. One bit after another using the same wire. Plotters and mice usually use this type of interface.

simplex One direction communication system (e.g. radio, TV).

simulation A process to generate test conditions that approximate real conditions, e.g. the use of flight simulators for pilot training.

software This is the set of programs, and procedures and related documentation associated with a computer program, compiled to perform a specific task.

spatial data Computer readable information on location of geometric objects represented in the computer.

stereoscopic Imparting a 3D effect; each eye receives slightly different image so that viewed together the image appears to have depth.

synthetic environments *See* virtual environment.

tactile feedback Feedback directed through or simulating the sense of touch or physical feel.

teleoperation Performing actions or manipulating objects, e.g. equipment, via robot or telepresence; sometimes referred to as telemanipulation.

telepresence Term coined by M. Minsky; 'remote' presence; sense of being present physically within remote scene resulting from recreation of the scene via computer; in virtual reality, a psychological experience resulting from immersion in virtual worlds.

texture mapping Filling in polygons with stored textured patterns; substituting surface effects for single colours instead of patterns to save computer resources or reduce processing time.

three dimensional *See* 3D.

virtual environment (VE) Space in which a user of VR technology imagines himself or herself and in which an interaction takes place; computer-generated world or process.

virtual prototype A realisation of an intended design or product to illustrate the characteristics of the product or design to users before actual construction; usually used as an exploratory tool for developers or communication prop for reviewing proposed design.

virus A computer virus is a particular sort of program that is introduced into a system to perform various functions, such as the destruction of programs and data, distortion of computer operation or espionage for particular information (such as passwords).

visualisation (VR) Presenting data in 'a' more comprehensible way in order to enhance its meaning and interpretation.

VR Virtual reality – also referred to as artificial reality; a computer-generated, digital, model of an environment that is highly interactive and attempts to eliminate separation between the user and machine.

workstation A single-user microcomputer, usually with high resolution graphics and high-speed processing capacity, capable of running applications independently or with other computers via a network. Workstations are generally considered more powerful than personal computers (PCs) although high-end PCs now match low-end workstations.

Index

2D drawing, 115
3D modelling, 116
abstraction, object-orientation, 48
access systems, 12
activity levels, 87
activity sequences, 88
Adjei-Kumi, T., 93
administration of databases, 25
advanced packages, project-management
    software, 77
agreed visions, 165–166
Ahuja, H. N., 120
AI *see* artificial intelligence
Alshawi, M., 130
alternative future construction models, 152
analysers, 147–148
animation, 28–29
Ansoff, J., 130
Anumba, C., 104, 112, 113
Aouad, G., 5, 43, 130
applications software, 7
AR *see* augmented reality
Artemis 7000, 80
artificial intelligence (AI), 45–46
Association for Information Management
    (Aslib), 59
   Aslib On-line Series, 59
attributes, drawing entities, 117
Au, T., 73, 83
augmented reality (AR), 36–40, 95–99
augmented virtuality (AV), 36–38
autoCAD software, 163
AV *see* augmented virtuality
Azuma, R., 36, 39

background data, 42–43
Bakos, 141

Baldwin, A, 113
Balfour Beatty, 93
Banxia software, 153
Barbour, 57
Barbour Index, 57
Barr, G., 75
base levels, project-management software,
    75–76
Basoz, N., 43
BCIS on-line (Building Cost Information
    Service), 59
Betts, M., 27, 68, 104, 127
BICC plc, 94
Bingham, Dr Peter, 159
Bjornsson, H., 136–137
blessed potential, 134–135
Boeckh (standard cost elements), 120
BOT/BOOT (build–operate–(own)–transfer),
    67
Brandon, P., 27, 104, 125
Brandon, P. S., 120
Bridges, A., 48
Bridgewater, C., 93
Brightman, J., 155
British Airways, 66
British Computer Association, 161
Brochner, J., 139
Brodie, K. W., 27
budgets, *91*, 163–164
Building Cost Information Service, 120
building process characterisation, 106
building products, 105, *107*
building-element relationships, 110–111
bulletin boards, 59
Burns, P., 43
bus structures, *10*, 11
Business Portfolio, 141–142

CAD *see* computer-aided design
CADD *see* Computer-aided design and Drafting
CAFE *see* Construction Alternative Futures Explorer
Callahan, M. T., 73
cameras, 31
Campbell, W. J., 120
Cannon, J., 152
Capron, H. L., 7
Cargil, C., 45
case base of methods, 119
case contents, 119
case-based reasoning (CBR), 47, 118–119
CAWS *see* Common Arrangement of Works Section
CBR *see* case-based reasoning
CCTV *see* closed-circuit television systems
CD-ROMs (Compact Disk, Read Only Memory), 22
central concept models, 154
CERN (Centre Européen de Recherches Nucléaires), 12
Cheshbrough, H. W., 66
Cheshire County Council, 159–162
CHESHIRElink, 161, 162
Chorafas, D., 23, 26
CICA *see* Construction Industry Computing Association
CIM *see* Common Information Model
CITE (Construction Industry Trading Electronically), 113
claim management, 75
client approach to planning and management, 159–162
client-server concept, 13
closed-circuit television systems (CCTV), 93
Co-ordinated Project Information (CPI), 56
Collier, J., 93
colour graphics, 152
Common Arrangement of Works Section (CAWS), 56–57
Common Information Model (CIM), 157
Common Technical Architecture (CTA), 166
communications
    computer, 7–9
    integrated systems, 170171
    technology, 158
    video conferencing, 29–31, *30*
company specific data, 42–43, 44
competitive advantage, 129–135
compose-on-screen facilities, 80

computer-aided design (CAD), 143
    databases, 23–24
    drafting, 115–117
    drawing, 117
    information transfer, 117–118
    preliminary designs, 113–115
Computer-aided design and Drafting (CADD), 18, 125
Computer-Readable Databases, 58
computers
    animation, 28–29
    applications, 59
    building modelling, 104–111
    communications, 79
    concurrent engineering, 112–121
    configurations, *4*
    crime, 68–69
    graphics, 17–18, 27–31
    information sources, 57–58
    integrated building design/construction, 104–111
    languages, 6–7
    project planning/scheduling, 73–101
    screens, 34
    software, 33–49
concurrent engineering, 112–121, 168–172
concurrent integration, 115
Conlin, J., 75, 81
Construction Alternative Futures Explorer (CAFE), 143, 153
construction firm strategy formulation, 146–155
Construction Forecasts and Research organisation, 55
Construction Industry Computing Association (CICA), 129, 133
construction industry model, 150–155
construction markets, 133, *134*
construction output values, *146*
construction strategies, 135–139, 146–155
construction-management functions, 107–110
consultant approach to design, 162–165
contacts (building-elements), 110–111
context data, 117
continuous inputs, 53
continuous simulations, 29, 84
contractors
    construction, 165–168
    project information sources, 57
copyright laws, 69
corporate information, 54–55

corporate intranets, 173, 175–176
cost-time forecasting, 85–92
costs, 14–15, *142*
   calculations, 89
   competitiveness, 142
   compliance, *91*
   estimates, 120
   leadership strategies, 140
   planning, 81–82
Cotton, B., 84
court cases, 93
Coventry, L., 29
Cox, J., 59
Coyne, R. D., 46
CPI *see* Co-ordinated Project Information
creditworthiness, 55
crime, 68–69
cross impact analysis, 149
cross-organisational use of information
   technology, 140
CSSP/Thomas Telford Software, 58
CST, 92
CTA *see* Common Technical Architecture
Cunningham, S., 29
current state of trade, 55
cyberspace, 33–49

data
   acquisition, 42–43, 115–117
   administration, 25
   collection structure, *151*
   construction management, 107–110
   information systems, 63–64
   input, 80
   integration, 82
   maintenance, 43
   models, 19–21, 43
   organisation, 52
   planning, 26
   processing, 64–65
   representation, *19*, 42–43
   storage, 17
   transfer nodes, *8*
   visualisation, 28
Data Coordination Report, 56
data-link features, 82
Data-Star, 59
databases, 18–27
   administration, 25
   approach, *20*
   availability extensions, 171
   integration, 26–27

   management, 19, 25–26
   models, 19–21
   online, 59–60
Date, C. J., 25
Day, A., 55
decentralisation, 167
decision emulation, 90
Decision Explorer software, 154
decision support systems (DSS), 24–25
decision support tool construction, 151
defenders, 147
Dejoie, R., 68
delay management, 75
delivery times, 112, 167
demand variability, 146
DeOliveira, 41
Department of Civil Engineering, 75
departmentalism, 160
design information access, 114
designers project information sources, 57
deterministic simulation models, 84
DIALOG Information Services, 60
digitising, 43
direct labour vs. sub-contraction costs, *142*
Directory of On-line databases, 59
Directory of Portable Databases, 58
discrete simulations, 29
distributed databases, 24
Distributed Management Task Force
   (DTMF), 157–158
dmtf.org, 158
domains, augmented reality, 39–40
DOS (Disc Operating System), 6
drawing entities, 117
Drogemuller, J., 113
DSL (Digital Subscriber Line), 9
DSS *see* decision support systems
DTMF *see* Distributed Management Task
   Force
Dulaimi, M. F., 66
Dunlop, C., 68
duration compliance, *91*
Dym, C. L., 46, 83

e-mail, 11
Earl, M., 127
Earnshaw, 27
East, W. E., 81
Eden, C., 149
EDICON (Electronic Data Interchange
   Construction), 113
EDM (Electronic Data Management), 113

Index

Electronic Information Series, 59
electronic networks, 59–60
element levels, 87
Elmasri, R., 18, 26
employment issues, 67–68, *146*
emulation, 90
encapsulation, 48
end user databases, 24
engineering databases, 23–24
ergonomical engineering, 68
ethical issues, 67–68
Evbuomwan, N., 104, 112, 113
events, 152
expert systems, 46–47, 83, 101
exploration models, 154–155
external databases, 24
external information, 53
extranets, 15, 175–176

Federal Energy Regulatory Commission, 174
Feiner, S., 39
Fentress Bradburn Architects Ltd, 175–176
Fenves, S., 112
Ferry, D. J, 120
file transfer protocols (FTP), 12
firewalls, 15–16
first-movers, 138–139
Fischer, G., 149
Fischer, M., 92, 104
fish tank systems, 39
Flaaten, P., 127
Flood, I., 49
Foley, J., 17
Ford Motor Company, 14
forecasts, 55, 85–92, 149, 154
foundation research, 151
frame-by-frame, 28
fraud, 68
FTP *see* file transfer protocols
function groups, 24–25
futurology, 149–150

Gale Research, 59
Gallacher, J., 126
GANs *see* global area networks
Garrett, J., 39
gcn.net/arena, 66
general data functions, 107, 110
generations (computer languages), 6–7
generic strategies, *128*
geographic information systems (GIS), 17,
    41–45

geometric modelling techniques, 104
Gibson, W, 33
GIS *see* geographic information systems
global area networks (GANs), 11, 65
Global Positioning Systems (GPSs), 44–45
global scenarios, 149
globalisation, 65–69, 174–175
glossary, 177–183
gloves, 35
Gonzales, A., 133–134, 135
government data, 54
GPSs *see* Global Positioning Systems
Gray, C., 84
Green, L., 84
Gurton, A., 14
Guss, C., 66, 67

Halpin, D., 84
Hampson, K., 130–131
Hannus, M., 15
hardware, 3–5, *4*
    information systems, 64
    virtual reality, 35
Harr, C., 15
Harris, F., 73
Hay Management Consultants, 161
Hay, R., 93
head-mounted displays (HMDs), 35, 38–39
headline information, 154
headphones, 31, 35
health problems (computer usage), 68
Hendrickson, C., 73, 83
Hering, J., 45
hierarchical models, 20, *21*
Hillebrandt, P. M., 152
Hills, M., 14
historical account of performance, 55
HMDs *see* head-mounted displays
Holloway, R. L., 39
horizontal integration, 114–115
hosts for networks, 59
Housing and Construction Statistics, 54
HTML (Hypertext Markup Language), 12
hub structures, *10*, 11
human factor engineering, 68
human resources, 63, 161
human-computer interfaces, 34–36
Hutton, G. H., 140
hybrid simulations, 29
hybrid virtual reality, 36, 95–99
hypothetical scenarios, 149

IAI *see* International Alliance for
Interoperability
iaiweb.lbl.gov, 158
ICA Flour Daniel, 174–175
ICE *see* Institution of Civil Engineers
ICONDA (the CIB International
Construction Databases), 59
IFC *see* Industry Foundation Classes
Illingworth, J. R., 73
images
databases, 22, 95
superimposing, 98, *100*
Imagina, 36
immersive virtual reality, 35, *35*
indexing cases, 119
industrial scenarios, 149
industry alliances, 158
Industry Foundation Classes (IFC), 104, 157,
158
information
characteristics, 60, 61–62
communication rules, 62
contents, 127
environments, 150
globalisation, 65–69
levels, 60–62
management, 61
processing, 64–65
sources, 53–60
systems, 24–25, 62–64, *63*, 159–162
transfer, 117–118
warehouse databases, 24–25
Information Sources, 58
Information Sources in Engineering, 58
Information Systems Strategy Unit,
159–162
information technology (IT), 125–143
application stages, *126–127*
budgeting, 163–164
competitive advantage, 129–135
construction strategies, 135–139,
146–155
current use, 125–129
first-movers, 138–139
implementation case studies, 159–173
Internet, 173–176
investments, 133
lead sustainability, 138
market networks, 139–140
planning, 163–164
strategic context, *128*
strategic management, 130, *131*

strategic opportunities, *129*
strategy formulation, 146–155
inheritance (object-orientation), 48
innergy.com/bin/newcashe?item = 395, 176
input project descriptions, 88
Institution of Civil Engineers (ICE), 159
Seminar, London (1994), 162, 165
Institution of Structural Engineers, 168
integration
communications, 170–171
computer-aided design/construction,
104–111
data, 82
databases, 26–27
images, 95
interfaces
human-computer, 34–36
knowledge-based systems, 47
structural design, 118–119
internal information, 53
internal strategies, 140–141
International Alliance for Interoperability
(IAI), 158
International Conference on Concurrent
Engineering, 168
Internet, 12–13, 173–176
interoperability, 48
intersection characterisation, 110–111
Intranet Design, 175–176
intranets, 13–16, 173–177
intuitive logic, 149, 150
invasion of privacy, 68
investments, 133, 163–164
ISDN (Integrated Services Digital Network),
9, 163
IT *see* information technology
item levels, 87

James Martin Associates, 130
JANET (Joint Academic NETwork), 59
Jenkin, F. P., 148
JFC
corporate information, 55
forecasts, 154
JIT *see* just-in-time
'jockeying for position', 127, *129*
John Laing, 165–168
Jones, D. R., 165
just-in-time (JIT) methods, 143

Kalawsky, R. S., 40
Kalay, Y. E., 104

Kartam, N., 49
Keen, P., 135
keyboards, 34
Keys, F., 14
Kirby, J. G., 81
Kishinio, F., 36
Kling, R, 68
knowledge-based expert systems (KBESs),
    46–47, 83, 101
knowledge-based simulation, 85–92, *86*
knowledge-based systems (KBSs), 46–47
Krueger, M., 33
Kumar, B., 46, 104, 112, 113, 115, 119
    artificial intelligence, 46
    building representation, 104
    concurrent engineering, 112, 113, 115,
    119
Kvaerner Oil & Gas Ltd, 168–172

Langford, D. A., 58, 67, 149
Langford, V., 55
languages (computer), 6–7
Lanier, J., 33
LANs *see* local area networks
Larijani, C. L., 33, 34
lead sustainability, 138
Legg, S., 23, 26
Leibich, T., 104
Levine, H. A., 75, 76
Levitt, R. E., 46, 83
Linux operating system, 6
Livelink, 173, 174
local area networks (LANs), 9–11
location groups, 24
long distance communications, 8–9
long-term strategies, 166
Loughborough University, 94
Lucas Management Systems, 80
Lundegard, R., 136–137

McCaffer, R., 73
Machover, C., 33
Mackenzie, John, 173
McLellan, R., 75
MacLeod, M., 29, 31
McLintock, 133
McManus, J. V., 168
macros, 82
Mair, G., 96
Male, S. P., 147, 149
management
    client approach, 159–162

databases, 24–25, 25–26
    geographic information system tools, 45
Management Information Systems (MIS),
    140–141
managerial-information, designers, 120
manipulation functions, 35
Mansfield, N. R., 75
many-to-many conferencing, 31
maps (models), 152
'Mark 1' Strathclyde Telepresence System,
    96
markets
    construction, *133*, *134*
    entry strategies, 138–139
    forecasts, 154
    networks, 139–140
Marston, V., 85, 92
Marwick, 133
mass-market levels, 76–77
master-plan generation, 120
mathematical simulations, 84
MCUs (multi-conferencing units), 31
Mead, P., 15
Means (R.S. Means Company), 58
Means (standard cost elements), 120
method-based representation, 119
mice, 34
microphones, 31
Miles, R. E., 147
Milgram, P., 36, 38
Millar, M., 126, 127, 137, 140
Mingus, P., 42, 45
MIS *see* Management Information Systems
mixed reality (MR), 38
MMHH *see* multimedia hard hat
mobile telecommunications networks, 65
mobile video telecommunications, 95
Model Computer Crime Act of American
    Data Processing Management
    Association, 69
models, 83–93
    alternative future construction, 152
    central concepts, 154
    computer-aided building, 104–111
    construction industry, 150–155
    data, 43
    databases, 19–21
    exploration, 154–155
    networks, 20, *21*
    stochastic simulation, 84
modems (Modulator-deModulator), 9
MR *see* mixed reality

Mraovic, B., 129
multidisciplinary concurrent engineering, 112–121
multidisciplinary design processes, *116*
multidisciplinary scenarios, 149
multimedia databases, 23
multimedia hard hat (MMHH), 94

National Buildings Specifications, 56
national scenarios, 149
Navathe, S., 18, 26
navigation (virtual reality), 35
Navon, R., 41
Networking protocols, 81
networks, 9–11
    cards, 31
    hosts, 59
    markets, 139–140
    models, 20, *21*
    neural, 48–49, *49*
    topology, *10*, 11
neural networks, 48–49, *49*
New Earnings Surveys, government data, 54
Newcombe, R., 53
non-immersive virtual reality, 36
North Sea oilfields, 169

object-orientation, 20, *21*, 33, 47–48
O'Brien, J. A., 25, 62
occasional information, 53
O'Conner, N., 36
off-line information sources, 57–59
offshore platforms, 169
Oliver, R., 84
Oliviera, L., 93
Oman, C., 42
on-line databases, 59–60
on-line hosts, 59–60
on-line information sources, 57–58, 59–60
one-to-many communications, 29
one-to-one meetings, 29
OpenPlan, 80
operating systems, 5–6
operational databases, 24
operational information, 60
organisation involvement, 166
organisation strategic responses, 147
OS/2 operating system, 6

parallel data transfer nodes, *8*
Parker, D., 42, 45
partnering concepts, 140

partnerships, 166–167
Paulson, B. C., 46
Peat, 133
penetration, 110–111
performance records, 55
PERINORM, 58
periodic information, 53
Perron, T. D., 7
PEST analysis, 54
photogrammetry, 17
pictorial synthesis, 17
Pidd, M., 84
Pilcher, R., 73
Pimental, K., 34, 40
planning and management, client approach, 159–162
planning processes, 73–101
Port, S., 113
Porter, M., 126, 127, *128*, 137, 140
post-mortem analysis, *92*
Powell, J., 93
Prasad, B., 112
preliminary design, 113–115
presentation graphics, 27
Price, A. D. F., 5
Price Adjustment Formula, 56
privacy issues, 68
processes
    information contents, 127
    modelling, 106
    representation, 105107
    visualisation, 28
procurement techniques, 66–67
production
    costs, 142
    planning, 120
    resources, 142
Production Drawings Code, 56
products
    information contents, 127
    models, 106
    representation, 105–107
programs, 59
Project Specification Code, 56
project-management software, 7483
    advanced packages, 77
    base levels, 75–76
    features, *78–79*, 80–82
    mass-market levels, 76–77
    sophisticated packages, 77–78
    specialist features, 75
    technical features, 75

projection (virtual reality), *36, 37*
projects
  based information, 55–57
  data, 107, 110
  information, *108–109*, 174–175
  levels, 86
  planning, 73–101
  schedules, 73–101, 169, *169*
  tracking, 80–81
Property Services Agency General
  Specification, 56
prospectors, 147
Prospero, 174–175

Ramo, J. C., 69
raster formats, 17, 43
Ray, C. F., 42
reactors, 147–148
'ready-to-use' design products, 120
real-time animation, 28–29
reality-virtuality (RV), 36, *39*
regional market differences, 147
relational database models, 20, *21*
remote sensing, 44–45, 95–96
reports (project-management software), 82
research programmes model, 150
resource planning, 81
resource-times-rate approach, 81–82
results presentation, 90–92
Retik, A.
  augmented reality, 39, 41
  building representation, 105, 106, 110,
    111
  computerised databases, 58
  concurrent engineering, 113, 115, 120
  project planning, 75, 81, 85, 92, 93, 94,
    96
retrieving cases, 119
RIBA *see* Royal Institute of British
  Architects
Ribarsky, W., 40, 93
RIBA.ti, 58
Rich, E., 46
ring structures, *10, 11*
Robinson, S., 84
Roorda, J., 43
Royal Institute of British Architects (RIBA),
  129
Rutter, P., 162
RV *see* reality-virtuality

satellites, remote sensing, 44–45

Saur-Bowker, 58
Scan-A-Bid, 59
scenarios, 149–155
  maps, 152, 153
  studies, 151–155
scheduling processes, 73–101
Schmitz, J., 65
scientific visualisation, 27
Scott, P. A. R., 44
Scott Wilson Kirkpatrick (SWK) group, 162
screens, 34
SDLC *see* systems development life cycle
search facilities, 58
security, 15–16
sequence establishment, 88
serial data transfer nodes, *8*
servers, 13
Shapira, A.
  building representation, 105, 110, 111
  virtual reality, 41, 93
shared visions, 165–166
She, T. H., 43
shells (software), 46–47
Simmons, G., 44
Simon, H. A., 119
simulations, 29, 83–93
  continuous, 29, 84
  deterministic methods, 84
  knowledge-based, 88, *89*
  mathematical, 84
  site environments, 40, 41
  statistical, 84
  steady-state models, 84
  stochastic models, 84
  virtual reality, 92–101
site environment simulation, 40
site management decision emulation, 90
Siyan, K., 15
Sleurink, H., 28, 29
Smailagic, A., 39
Smith, I., 39
SMM7 *see* Standard Method of
  Measurement
Snow, C. C., 147
social issues, 67–68
software, 5–9
  characteristics, 74–75
  information systems, 64
  piracy, 69
  technical applications, 33–49
sophisticated packages, project-management
  software, 77–78

spatial indexes, 45
specific data, 42–43
Stalker, R., 39
standard cost elements, 120
Standard Method of Measurement, 56
Standard Method of Measurement 7 (SMM7), 57
Stanford University, 43
star structures, *10*, 11
statistical simulations, 84
steady-state simulation models, 84
STEP, 158
Stiny, G., 114
stochastic simulation models, 84
Stocks, R. K., 147
Stockyk, J., 15
strategic context (information technology), *128*
strategic fits, *132*
strategic information, 60
strategic management
    incorporation, 136
    information technology, 130, *131*, 143
    principles, 155
    turbulent environments, 148150
strategic opportunities, *129*
strategic planning, 148–150
strategic responses, 147
strategic system development, 120–121
strategic treasure, 135
strategy formulations, 146–155
strath.ac.uk/Departments/Civveng/conman/vccsrg/, 93
Strathclyde Telepresence System, 96
Strathclyde University, 93, 94–102, 143, 148
structural design, 118–119
structures
    data collection, *151*
    databases, 18
Stuart, A., 14
sub-contraction vs. direct labour costs, *142*
superimposing images, 98, *100*
Sviden, O., 148
Swartz, P., 149
SWK *see* Scott Wilson Kirkpatrick
system development strategies, 120–121
system operation knowledge-based simulation, 88, *89*
systems development life cycle (SDLC), 7

tactical information, 60
technology

construction market requirements, *133*, *134*
    strategies, *132*
Technology Foresight Programme, 149
TED (Tenders Electronic Daily), 59
Teece, D. J., 66
Teicholz, P., 104
Teixeira, K., 34, 40
telecommunication, 93–102
telepresence (TP), 95–99
    mobile version, *98–99*
    static version, 97
text databases, 22
Thomas, D., 42, 43
Thorpe, S., 94
Tice, S., 33
time estimates, 120
time-based construction claims, 80–81
Tom, P. L., 25
Tommelein, I., 84
tools, visualisation, 27
topology, *10*, 11
Topping, B., 49
Torvalds, Linus, 6
total quality management (TQM), 143
traditional databases, 22
training (virtual reality), 40, 41
trans-organisational, information technology, 140
TransCanada Pipelines Ltd, 173–174
transnational business strategies, 65
Transparent Telepresence Research Group, 96
Treacy, M., 141
tree structures, *10*, 11
trend impact analysis, 149
turbulent environments, 145150
turnkey methods
    contractors, 140
    shape design product, 120

University of Loughborough, 94
University of Stanford, 43
University of Strathclyde, 143, 148
    Department of Civil Engineering, 75
    Virtual Construction Stimulation Research Group, 93, 94–102
UNIX operating system, 6
unnecessary investments, 134
unwanted necessities, 134
user interfaces, 47

vague scenarios, 149
van der Heiijden, K., 149, 150
variability of demand, 146
VAX computer systems, 163
VE *see* virtual environment
vector format, 43
Vector Pipeline, 174
vertical integration, 114
video cards, 31
videoconferencing, 29–31, *30*, 93
Vince, J., 36
Vincent, S. P. R., 44
Virtual Construction Stimulation Research
    Group, 93, 94–102
virtual environment (VE), 34
virtual organisations, 65–67, 141
Virtual Reality Modelling Language
    (VRML), 12–13
virtual reality (VR), 33–49, 92–101
viruses, 69
visual monitoring, *96*
visualisation techniques, 27–29, 92–101
volume visualisation, 28
VR *see* virtual reality
VRML *see* Virtual Reality Modelling
    Language

Wack, P., 148
WANs *see* wide area networks
Ward, J., 136
Warszawski, A., 106, 120
Wayment, M., 150
WBEM (Web-Based Enterprise
    (Management), 157
web browsers, 175–176
Welcome Software Technology, 80
Wessex (Wessex Software UK Ltd), 58–59
Whyte, J., 93
wide area networks (WANs), 11
Wilkinson, 173–174
Williamson, O., 141–142
Windows operating system, 6
Wiseman, 27
Wix, J., 104
Wix, Y., 47
working conditions, 67–68, 161
works information, 56
World Wide Web (WWW), 12–13
worms, 69
www *see* individual names; World Wide
    Web

XML (eXtensible Markup Language), 12

Zouein, P., 84